CAMBRIDGE LIBRARY COLLECTION

Books of enduring scholarly value

Darwin

Two hundred years after his birth and 150 years after the publication of 'On the Origin of Species', Charles Darwin and his theories are still the focus of worldwide attention. This series offers not only works by Darwin, but also the writings of his mentors in Cambridge and elsewhere, and a survey of the impassioned scientific, philosophical and theological debates sparked by his 'dangerous idea'.

Physical Geography

Mary Somerville (1780–1872) would have been a remarkable woman in any age, but as an acknowledged leading mathematician and astronomer at a time when the education of most women was extremely restricted, her achievement was extraordinary. Laplace famously told her that 'There have been only three women who have understood me. These are yourself, Mrs Somerville, Caroline Herschel and a Mrs Greig of whom I know nothing.' Mary Somerville was in fact Mrs Greig. After (as she herself said) translating Laplace's work 'from algebra into common language', she wrote On the Connexion of the Physical Sciences (1834), also reissued in this series. Her next book, the two-volume Physical Geography (1848), was a synthesis of geography, geology, botany and zoology, drawing on the most recent discoveries in all these fields to present an overview of current understanding of the natural world and the Earth's place in the universe.

Cambridge University Press has long been a pioneer in the reissuing of out-of-print titles from its own backlist, producing digital reprints of books that are still sought after by scholars and students but could not be reprinted economically using traditional technology. The Cambridge Library Collection extends this activity to a wider range of books which are still of importance to researchers and professionals, either for the source material they contain, or as landmarks in the history of their academic discipline.

Drawing from the world-renowned collections in the Cambridge University Library, and guided by the advice of experts in each subject area, Cambridge University Press is using state-of-the-art scanning machines in its own Printing House to capture the content of each book selected for inclusion. The files are processed to give a consistently clear, crisp image, and the books finished to the high quality standard for which the Press is recognised around the world. The latest print-on-demand technology ensures that the books will remain available indefinitely, and that orders for single or multiple copies can quickly be supplied.

The Cambridge Library Collection will bring back to life books of enduring scholarly value (including out-of-copyright works originally issued by other publishers) across a wide range of disciplines in the humanities and social sciences and in science and technology.

Physical Geography

VOLUME 2

MARY SOMERVILLE

CAMBRIDGE UNIVERSITY PRESS

Cambridge, New York, Melbourne, Madrid, Cape Town, Singapore,
São Paolo, Delhi, Dubai, Tokyo

Published in the United States of America by Cambridge University Press, New York

www.cambridge.org
Information on this title: www.cambridge.org/9781108005210

This edition first published 1848
This digitally printed version 2009

ISBN 978-1-108-00521-0 Paperback

PHYSICAL GEOGRAPHY.

BY

MARY SOMERVILLE,

AUTHOR OF THE "CONNEXION OF THE PHYSICAL SCIENCES,"
"MECHANISM OF THE HEAVENS."

IN TWO VOLUMES.

VOL. II.

WITH A PORTRAIT.

LONDON:
JOHN MURRAY, ALBEMARLE STREET.

1848.

CONTENTS OF VOL. II.

CHAPTER XXIV.

CHAPTER XXV.

CHAPTER XXVI.

CHAPTER XXVII.

CHAPTER XXVIII.

CHAPTER XXIX.

CHAPTER XXX.

CHAPTER XXXI.

PHYSICAL GEOGRAPHY.

CHAPTER XVIII.

RIVER SYSTEMS OF NORTH AMERICA—RIVERS OF CENTRAL AMERICA—RIVERS OF SOUTH AMERICA, AND OF AUSTRALIA.

North America is divided into four distinct water systems by the Rocky Mountains, the Alleghannies, and a table-land which contains the great lakes, and separates the rivers that flow into the Arctic Ocean from those that go to the Gulf of Mexico. This table-land is a level, nowhere more than 1200 or 1500 feet above the surface of the sea, and is itself drained by the St. Lawrence and the rivers that flow into Hudson's Bay. The St. Lawrence rises in Lake Superior, and, after joining the five great lakes, runs north-east into the Atlantic, and ends in a wide estuary. It has a basin of 537,000 square miles, of which 149,000 are covered with water, exclusive of the many lesser lakes with which it is in communication.

North of the watershed there is an endless and intricate labyrinth of lakes and rivers, almost all connected with one another. But the principal streams of these arctic lands are the Great Fish River, which flows north-east in a continued series of dangerous and all-but impassable rapids to the Arctic Ocean at Melville Strait. The Copper-mine River, of much the same character, after traversing many lakes, enters the Icy Sea at George IV.'s Gulf; and the M'Kenzie, a stream of greater magnitude, formed by the confluence of the Peace River and the Athabasca from the Rocky Mountains, after flowing north over 16 degrees of latitude, enters the frozen ocean in the Esquimaux country somewhere beyond the arctic circle. All these rivers are frozen more than half the year, and the M'Kenzie, in consequence of its length and direction from south to north, is subject to floods like the Siberian rivers, because its lower course remains frozen for several hundred miles, long after the upper part is thawed, and the water, finding no outlet, flows over the ice and inundates the plains.

South of the table-land the valley of the Mississippi extends for 1000 miles, and the greatest of North American rivers has its origin in the junction of streams from the small lakes Itaska and Ussawa, on the table-land, at no greater height than 1500 feet above the sea. Before their junction these streams frequently spread out into sheets of water, and the Mississippi does the same in the upper part of its course. This river flows from north to south through more degrees of latitude than any other, and receives so many tributaries of the higher orders, that it would be difficult even to name them. Among those that swell its volume from the Rocky Mountains, the Missouri, the Arkansas, and the Red River are the largest, each being in itself a mighty stream, receiving tributaries without number. Before their junction the Missouri is a much superior stream, both in length and volume, to the Mississippi, and has various affluents larger than the Rhine. It rises in about 44° N. lat., and runs partly in a longitudinal valley of the Rocky Mountains and partly at their foot, and drains the whole of the country on the right bank of the Mississippi, between the 49th and 40th parallels of north latitude. It falls in cataracts through the mountain regions, but in the plains it sometimes passes through dense forests and sometimes through large prairies, in all accomplishing 3000 miles in a very tortuous and generally south-eastern direction, till it is confluent with the Mississippi near the town of St. Louis. Lower down the Mississippi is joined first by the Arkansas, 2000 miles long, with many accessories, and then by the Red River, the former from the Rocky Mountains, but the latter, which rises in the table-land of New Mexico, is fed by streams from the Sierra dal Sacramento, and enters the main stream not far from the beginning of the delta which stretches in a long tongue of land into the Gulf of Mexico.

The tributaries from the Rocky Mountains, though much longer, run through countries of less promise than those which are traversed by the Ohio and the other rivers that flow into the Mississippi on the east, which offer advantages unrivalled even in this wonderful country, only beginning to be developed.

The Ohio is formed by the union of the rivers Alleghanny and Monongahela, the latter from the Laurel ridge of the Alleghanny chain in Virginia, the former comes from sources near Lake Erie, and the two unite at Pittsburg, from whence the river winds for 948 miles through some of the finest States of the Union, till its junction with the Mississippi, having received many accessories, six of which are navigable streams. There are some obstacles to navigation in the Ohio, but they have been avoided by canals. Other canals join both the Mississippi and its branches with Lake Erie, so that there is an internal water communication between the St. Lawrence and the Gulf of Mexico. The whole length of the Mississippi is 3160 miles, but if the Missouri be considered the main stem, it is 4265, and the joint stream drains an area of about a million and a quarter of square miles. The breadth of the river nowhere corresponds with its length. At the confluence of the Missouri each river is half a mile wide, and after the junction of the Ohio it is not more. The depth is 168 feet where it enters the Gulf of Mexico at New Orleans. This great river is a rapid, desolating torrent loaded with mud: its violent floods, by the melting of the snow in the high latitudes, sweep away whole forests, by which the navigation is rendered very dangerous; and the trees, being matted together in masses many yards thick, are carried down by the spring floods, and deposited over the delta and Gulf of Mexico for hundreds of square miles.

North America can boast of two other great water systems, one from the eastern versant of the Alleghannies, which flows into the Atlantic, and another from the western versant of the Rocky Mountains, which runs into the Pacific.

All the streams that flow eastward through the United States to the Atlantic are short and comparatively small, but of the highest utility, because many of them, especially those to the north, end in gulfs of vast magnitude, and the whole are so united by canals, that few places are not accessible by water, one of the greatest advantages a country can possess. There are at least 24 canals in the United States, the whole length of which is 3101 miles.

Many of the streams that ultimately come to the Atlantic rise in the western ridges of the Alleghanny chain, and traverse its longitudinal valleys before leaving the mountains to cross the Atlantic slope, which terminates in a precipitous ledge for 300 miles parallel to the range. By falling over this rocky barrier in long rapids and picturesque cascades, they afford an enormous and extensive water power: and as the rivers are navigable from the Atlantic quite across the maritime plains, these two circumstances have determined the location of most of the principal cities of the United States at the foot of this rocky ledge, which, though not more than 300 feet high, has had a greater influence on the political and commercial interests of the Union, than the highest chains of mountains have had in other countries.

The watershed of the Rocky Mountains lies at a greater distance from the Pacific than that of the Alleghannies from the Atlantic; consequently the rivers are longer, but they are few and little known. The largest are the Oregon or Colombia and the Rio Colorado. The former has its source not far from those of the Rio del Norte, and after an exceedingly tortuous course, in which it receives many tributaries, it falls into the Pacific at Astoria. The Colorado is a Mexican stream, which comes from the Sierra Verde, and falls into the Gulf of California.

There are many streams in Central America, and above ten rivers that are navigable for some miles; six of these fall into the Gulf of Mexico and Caribbean Sea, and four into the Pacific.

The Andes, the extensive watershed of South America, are so close to the Pacific, that, excepting a few small streams at their southern extremity, there are no rivers on that side, and even the streams that rise in the western Cordilleras find their way to the eastern plains.

The Magdalena, at the northern end of the Andes, though a secondary river in America, is 620 miles long. It rises in the central chain, at the divergence of the Cordillera of Santa Fé de Bogota, and enters the Caribbean Sea by various channels, navigable to Honda. The Cauca, its only feeder on the left, comes from Popayan, and is nearly as large as its primary, to which it runs parallel the greater part of its course. Many streams join the Magdalena on the right, as the Funzha, which waters the elevated plain of Bogota and forms the cataract of Tequendama, one of the most beautiful and wildest scenes in the Andes. The river rushes through a chasm 30 feet wide, which appears to have been formed by an earthquake ; and at a double bound descends 530 feet into a dark gloomy pool, illuminated only at noon by a few feeble rays. A dense cloud of vapour rising from it is visible at the distance of 15 miles. At the top the vegetation is that of a temperate climate, while palms grow at the bottom.

With the exception of the Magdalena, all the water from the inexhaustible sources of the Andes is poured into the Orinoco, the river of the Amazons, and the Rio de la Plata, which convey it eastward across the continent to the Atlantic.

The basins of these three rivers are separated in their lower parts by the mountains and high lands of the Parima and Brazil; but the upper parts of the basins of all three, towards the foot of the Andes, form an extensive level, and are only divided from one another by imperceptible elevations in the plains, barely sufficient to form the watersheds between the tributaries of these majestic rivers. This peculiar structure is the cause of the natural canal of the Cassiquiare, which joins the upper Orinoco with the Rio Negro, a principal affluent of the Amazons. Ages hence, when the wilds are inhabited by civilized men, the tributaries of these three great rivers, many of which are navigable to the foot of the Andes, will, by means of canals, form a water system infinitely superior to any that now exists.

The Orinoco, altogether a Colombian river, rises in the Sierra del Parima, 200 miles east of Duida, and maintains a westerly course to San Fernando de Atabapa, where it receives the Atabapa, and Guaviare, which is larger than the Danube, and here ends the upper Orinoco. The river then forces a passage through the Sierra del Parima, and runs due north, for three degrees of latitude, between banks almost inaccessible; its bed is traversed by dykes and filled with boulders of granite, and islands clothed with a variety of magnificent palm-trees. Large portions of the river are here engulfed in crevices, forming subterranean cascades; and in this part are the celebrated falls of the Atures and Apures, 36 miles apart, which are heard at the distance of many miles. At the end of this tumultuous part of its course it is joined by the Apure, a very large river, and then runs eastward to its mouth, where it forms a large delta, and enters the Atlantic by many channels. As the upper Orinoco runs west, and the lower Orinoco east, it makes a complete circuit round the Parima mountains, so that its mouth is only two degrees distant from the meridian of its sources.

The Cassiquiare leaves the Orinoco at the point where the rapids begin, and joins the Rio Negro, a chief tributary of the Amazons, at the distance of 180 miles.

The Orinoco is navigable 1000 miles, and at all seasons; a fleet might ascend it from the Dragon's mouth to within 45 miles of Santa Fé de Bogota.

It receives many navigable rivers, of which the Guaviare, the Atures, and the Meta are each larger than the Danube. The Meta may be ascended to the foot of the Andes; its mean depth is 36 feet, and in many places 80 or 90. It rises so high in the Andes, that Baron Humboldt says the vegetable productions at its source differ as much from those at its confluence with the Orinoco, though in the same latitude, as the vegetation of France does from that of Senegal. The larger feeders of the Orinoco come from the Andes, though many descend to it from both sides of the Parima, in consequence of its long circuit among these mountains.

The basin of the Orinoco has an area of 300,000 square miles, of which the upper part is impenetrable forest, the lower is Llanos.

The floods of the Orinoco, like those of all rivers entirely within the torrid zone, are very regular, and attain their height nearly at the same time with those of the Ganges, the Niger, and the Gambia. They begin to swell about the 25th of March, and arrive at their full and begin to decrease on the 25th of August. The inundations are very great, owing to the quantity of rain that falls in the wooded regions, which exceeds 100 inches in a year.

Below the confluence of the Apure, the river is three miles and a quarter broad, but during the floods it is three times as much. By the confluence of four of its greatest tributaries, at the point at which it bends to the east, a low inland delta is formed, in consequence of which 3600 square miles of the plain are under water during the inundation. The Orinoco in many parts smells of musk, from the number of dead crocodiles.

Upper Peru is the cradle of the Amazons, the greatest of rivers. It issues in two streams from the Lauricocha or lake Laura, in the elevated plain of Bombon, on the summit of the Andes. Joined by many other streams, it pursues a northerly course between the lateral Cordilleras, till it bursts through the eastern ridge by the Pongo or pass of Manseriche, and descends to the flat and wooded plain at the foot of the mountains; from whence it flows uniformly eastward till it reaches the Atlantic, having accomplished a course of 3200 miles, or more properly 4000, including its windings, and drained an area of two millions and a half of square miles, which is ten times the size of France. In some places it is 600 feet deep; it is navigable 2200 miles from its source, and is 96 miles wide at its mouth.

The name of this river is three times changed in its course: it is known as the Marañon from its source to the confluence of the Ucayale; from that point to its junction with the Rio Negro, it is called the Solimoes; and from the Rio Negro till it enters the ocean, it is the river of the Amazons.

The number, length, and volume of its tributaries are in proportion to its magnitude, even the affluents of its affluents are noble streams. More than 20 superb rivers, navigable almost to their sources, pour their waters into it, and streams of less importance are numberless. Two of the largest, the Huallago and Ucayale, like their primary, rise in the plains of Bombon ; the former has its origin in the mining district of Pasco, and after a long northern course between the Cordilleras it breaks through a gorge similar to that of Manseriche, and joins the Maranon in the plains : it is almost a mile broad above its junction. The Spanish Governor of Peru sent Pedro de Ursoa down this river, in the year 1560, to search for the lake of Parima, and the city of El Dorado. The Ucayale, not inferior to the Marañon itself, rises 90 miles east of the city of Lima. In a course of 1080 miles it is fed by accessories from an enormous extent of country, and at its junction with the main stream, near the mission of San Joachim de Omaquas, a line of 50 fathoms does not reach the bottom. By these streams there is access to Peru, and there is communication between the Amazons and the most distant regions around by the other navigable feeders. On the south it is connected with Bolivia and Brazil, by the Beni ; and the Madiera, which is its greatest affluent, comes near the sources of the Paraguay, the principal accessory of the Rio de la Plata. The river of the Amazons is not less extensively connected on the north. The high lands of Colombia are accessible by the Putumoya, the Japura, and other great navigable rivers ; the Rio Negro, nearly nine miles broad a little way above its junction with the Amazons, unites the latter with the Orinoco by the Cassiquiare ; and lastly the sources of the Rio Branco come very near to those of the Essequibo, an independent river of Demerara.

The main stream, from its mouth, nearly throughout its length, is full of river islands, and most of its tributaries have deltoid branches at their junction with it. The annual floods of the Amazons are less regular than those of the Orinoco, and as the two rivers are in different hemispheres, they occur at opposite seasons. The Amazons begins to rise in December, is at its greatest height in March, and its least in July and August. The quantity of rain that falls in the deep forests traversed by this river is so great, that were it not for the enormous evaporation, and the streams that carry it off, the country would be flooded annually to the depth of eight feet. The Amazons is divided into two branches at its mouth, of which one joins the Parà, south of the island of Das Joanes, the other enters the ocean to the north of it.

The water of some of the rivers in equatorial America is white; in others it is of a deep coffee colour, or dark green, when seen in the shade, but perfectly transparent, and when ruffled by a breeze, of a vivid green, like some of the Swiss Lakes. In Scotland the brown waters come from peat mosses, but it is not so in America, as they occur as often in forests as in savannahs. Mr. Schomburgk thinks they are stained by the iron in the granite; however, the colouring matter has not been chemically ascertained. The Orinoco and the Cassiquiare are white; the

Rio Negro is black, as its name implies, yet the water does not stain the rocks, which are of a dazzling white. Black waters are sometimes, though rarely, found on the table-land of the Andes.

The Rio de la Plata forms the third great water system of South America. The Rio Grande, its principal stream, rises in the mountains of Minas Geraes, in Brazil, and runs 500 miles on the table-land from N. to S., before it takes the name of Paranà. For more than 100 miles it is a continued series of cataracts and rapids, the greatest of which is the Salta Grande, about 24° 5′ lat. Above the fall the river is three miles broad, when all at once it is confined in a rocky pass only 60 yards wide, through which it rushes over a ledge with a thunderous noise, heard at the distance of many miles. The Paranà receives three large rivers on the right; the Paraguay, the Pilcomayo, and the Vermejo, all generally tending to the south, unite at different distances before entering their primary at Corrientes. The Paraguay, 1200 miles long, is the finest of these; in its upper part it is singularly picturesque, adorned with palms and other tropical vegetation, and its channel islands are covered with orange groves. It springs from a chain of seven lakes, in the southern slopes of the Campos Parecis, in Brazil, and may be ascended by vessels of considerable burthen through 19 degrees of latitude. The Pilcomayo and Vermejo both come from Bolivia; the former traverses the desert of the Gran Chaco, the latter the district of Tarija. At Santa Fé the

La Plata turns eastward, and before entering the Atlantic is augmented by the Uruguay from the north, which takes its name from the turbulence of its streams.

The Rio de la Plata is 2700 miles long, and for 200 miles from its mouth, up to Buenos Ayres, it never is less than 170 miles broad. Were it not for the freshness of its water, it might be mistaken for the ocean; it is, however, shallow and loaded with mud.

The Paraguay is subject to dreadful floods; in 1812 the atmosphere was poisoned by the putrid carcasses of drowned animals; the ordinary annual inundations of the Paranà, the principal or upper branch of the La Plata, cover 36,000 square miles.

In consequence of the vast extent of the very level plains along the base of the Andes, the basins of the three great rivers are apparently united. So small are the elevations that determine their direction, that, with the exception of a portage of three miles, a vessel might sail from Buenos Ayres, in 35° S. lat., to the mouth of the Orinoco, in 9° N. lat., by inland navigation.

There are various rivers in South America unconnected with those described, which in any other country would be esteemed of a high order. Of many that descend from the mountains of Parima, the Essequibo is the largest, fed by the streams of Guiana. Its general width is a mile and a quarter; its water, though black, is transparent; and on its banks, and those of all its adjuncts, the forest reigns in impenetrable thickness.

The Parà and San Francisco are the chief Brazilian rivers; both rise on the table-land: the former results from the union of the Tocantins and Araguay; it descends from the high lands in rapids in its northerly course, and after running 1500 miles joins the southern branch of the Amazons before entering the Atlantic, south of the island Das Joanes. The San Francisco is only 1275 miles long, and after travelling northward between mountain ranges parallel to the coast, it breaks through them, and reaches the ocean about the 11th degree S. lat. As in the Appalachian chain, so here many little rivers come down the edge of the table-land to the level maritime plains of the Atlantic.

In the far south the Rio Negro, and some other streams from the Chilian Andes, run through, but do not fertilize, the desolate plains of Patagonia.

RIVERS OF NEW HOLLAND.

After America, the land of the river and the flood, New Holland appears in more than its usual aridity. The absence of large rivers is one of the greatest impediments to the improvement of this continent. What it may possess in the interior is not known, but it is certain that no large river discharges its water into the ocean, and most of the small ones are absorbed before they reach it.

The streams from the mountains on the eastern side of the continent are mere torrents, and would have short courses did they not run in longitudinal valleys, as, for example, the Hawkesbury. The Murumbigee, the Lachlan and McQuarrie, formed by the accumulation of mountain torrents, are the largest.

The Murumbigee rises in the ranges west of St. George's Lake, and running south-west, meets the Lachlan, of unknown origin, coming from the east. After their junction they pass through the Alexandrine Marsh, and run into the Murray, a much larger stream, though only 350 feet broad, and not more than 20 feet deep, and on entering the ocean in Endeavour Bay it is too shallow even for boats. The Darling is supposed to be merely the upper part of the Murray, probably rising towards the head of St. Vincent's Gulf. The origin of the Macquarrie is unknown; it is called the Fish River, between Bathurst and Sydney; after running 600 miles north-west, it is lost in the marshes.

Swan River, on the western side of the continent, has much the same character; and from that river to the Gulf of Carpentaria, along the whole of the western and northern shores of the continent, there are none. The want of water makes it hardly possible to explore the interior of this continent.

CHAPTER XIX.

LAKES — NORTHERN SYSTEM OF THE GREAT CONTINENT— MOUNTAIN SYSTEM OF THE SAME—AMERICAN LAKES.

The hollows formed on the surface of the earth by the ground sinking or rising, earthquakes, streams of lava, the intersection of strata, and those that occur along the edges of the different formations, are generally filled with water, and constitute systems of lakes, some salt and some fresh. Many of the former may be remnants of an ancient ocean left in the depression of its bed during its retreat as the continents arose.

Almost all lakes are fed by springs in their beds, and they are occasionally the sources of the largest rivers. Some neither receive tributaries nor have outlets: the greater number do both. The quantity of water in lakes varies with the seasons everywhere, especially from the melting snow on mountain chains and high latitudes, and between the tropics from periodical rains. Small lakes occur in mountain passes, formed by water which runs into them from the commanding peaks: they are frequently, as in the Alps, very transparent, of a bright green or azure hue. Large lakes are common on tablelands and in the valleys of mountainous countries, but the largest are on extensive plains. The basin of a lake comprehends all the land drained by it; consequently it is bounded by an imaginary line passing through the sources of all the waters that fall into it.

There are more lakes in high than in low latitudes, and in this respect there is a great analogy between the northern plains of the two principal continents. Sheets of water of great beauty occur in the mountain valleys of the British Islands, of Norway and Sweden, countries similar in geological structure; and besides these, there are two regions in the old world in which lakes particularly abound. One begins in the low coast of Holland, goes round the southern and eastern sides of the Baltic, often passing close to its shores, along the Gulf of Bothnia, and through the Siberian plains to Behring's Straits. The lakes which cover Finland, and the great lakes of Ladoga and Onega, lie in a parallel direction: they occupy transverse rents which had taken place across the palæozoic strata while rising in a direction from S.W. to N.E. between the Gulf of Finland and the White Sea; that elevation was perhaps also the cause of the cavities now occupied by these two seas. Ladoga is the largest lake in this zone, having a surface of nearly 1000 square miles. It receives tributary streams and sends off its superfluous water by rivers, and Onega does the same; but the multitude of small steppe lakes among the Ural Mountains and in the basin of the river Obi neither receive nor emit rivers, being for the most part mere ponds, though of great size, some of fresh, some of salt water, lying close together, a circumstance which has not been accounted for: the lakes in the low Siberian plains have the same character.

The second system of lakes in the old continent follows the zone of the mountain mass, and comprehends those of the Pyrenees, Alps, Apennines, Asia Minor, the Caspian, the lake of Aral, together with those on the table-land and in the mountains of central Asia.

In the Pyrenees lakes are most frequent on the French side: many are at such altitudes as to be perpetually frozen: one on Monte Perdido, 8393 feet above the sea, has the appearance of an ancient volcanic crater. There is scarcely a valley in the Alpine range and its offsets that has not a sheet of water, no doubt owing to the cavities formed during the elevation of the ridges, and in some instances to subsidence of the soil. There are more lakes on the north than on the south side of the Alps—the German valleys are full of them. In Bohemia, Galicia, and Moravia, there are no less than 30,000 sheets of water, besides great numbers throughout the Austrian empire.

Of the principal lakes on the northern side of the Alps, the lake of Geneva, or lake Leman, is the most beautiful, from its situation, the pure azure of its waters, and the sublime mountains that surround it. Its area, of about 100 square miles, is 1150 feet above the sea, and at Meillerie it is 1000 feet deep. The lake of Lucerne is 1400 feet above the sea,

c 2

and the lake of Brienz 1900 feet. The Italian Alpine lakes are at a lower level: the Lago Maggiore has only 678 feet of absolute altitude: they are larger than those in the north, and with the advantages of an Italian climate, sky, and vegetation, they surpass the others in beauty, though the mountains that surround them are less lofty.

These great lakes are fed by streams from glaciers in the higher Alps, and many large rivers issue from them. In this respect they differ from most of the lakes in lower Italy, which, with few exceptions, are craters of ancient volcanos, or perhaps ancient craters of elevation, where the earth had been swelled up by subterranean vapour without bursting, and had sunk down again into a hollow when the internal pressure was removed.

In Syria, the lake of Tiberias and the Dead Sea, sacred memorials to the Christian world, are situate in the deepest cavity on the earth. The surface of the former, 466 feet below the level of the Mediterranean, is adorned with verdant plains and aromatic shrubs; while the heavy, bitter waters of the Dead Sea, 1312 feet below the same level, is a scene of indescribable desolation and solitude, encompassed by desert sands, and bleak, stony salt-hills. Thus there is a difference of level of 1000 feet in little more than 60 miles, which makes the course of the Jordan very rapid. The water of the Dead Sea is so acrid that it irritates the skin; and as it contains 26·24 per cent. of chlorides, it is more buoyant, and contains a greater proportion of salt, than any that is known, except the small lake of Elton, east of the Volga.

Though extensive sheets of water exist in many parts of Asia Minor, especially in Bithynia, yet the characteristic feature of that country, and of all the table-land of Western Asia and the adjacent steppes, is the number and magnitude of the saline lakes. A region of salt lakes and marshes extends at least 200 miles along the northern foot of the Taurus range, on a very elevated part of the table-land of Anatolia. There are also many detached lakes, some exceedingly saline. Fish cannot live in the lake of Toozla, and if a bird dips its wings in the water, they are incrusted with salt on drying: it is shallow and subject to excessive evaporation. Neither can any animal exist in the lake of Shahee or Urmiah, on the confines of Persia and Armenia, 300 miles in circumference: its water is perfectly clear, and contains a fourth part of its weight of saline matter. These lakes are fed by springs, rain, and melted snow, and having no emissaries, the surplus water is carried off by evaporation.

It is possible that the volcanic soil of the table-land may be the cause of this exuberance of salt water; yet there are many fresh-water lakes in their immediate vicinity. Lake Van, a sheet of fresh water, 240 miles in circumference, is separated from the salt lake Urmiah only by a range of hills, and there are many other pieces of fresh water in that neighbourhood.

Persia is singularly destitute of water: the lake

of Zorah, on the frontiers of Afghanistan, having an area of 18 square miles, is the only piece of water on the western part of the table-land of Iran.

It is evident, from the saline nature of the soil and the shells it contains, that the plains round the Caspian, the lake of Aral, and the steppes, even to the Ural Mountains, had once formed part of the Black Sea. 57,000 square miles of that country are depressed below the level of the ocean, a depression which extends northwards beyond the town of Saratov, 300 miles distant from the Caspian. The surface of the Caspian itself, which is 83·6 feet below the level of the ocean, is its lowest part, and has an area of 18,000 square miles, nearly equal to the area of Spain. In Europe alone it drains an extent of 850,000 square miles, receiving the Volga, the Ural, and other great rivers on the north. It has no tide, and its navigation is dangerous from heavy gales, especially from the S.E., which drive the water miles over the land: a vessel was stranded 46 miles inland from the shore. It is 600 feet deep to the south, but is shallower to the east, where it is bounded by impassable swamps many miles broad. The lake Elton, on the steppe east of the Volga, has an area of 130 square miles, and furnishes two-thirds of the salt consumed in Russia. Its water yields 29·13 per cent. of solid matter, while the water of the Dead Sea has 26·24 per cent. of saline ingredients; but it contains sulphate of magnesia, whereas Lake Elton has chloride of calcium.

The lake of Aral, which is shallow, is higher than the Caspian, and has an area of 3372 square miles. It has its name from the number of small islands at its southern end, Aral signifying "island" in the Tartar language. Neither the Caspian nor the lake of Aral have any outlets; though they receive large rivers, they, and all the lakes in Persia, are decreasing in extent and becoming more salt, the quantity of water supplied by tributaries being less than that lost by evaporation. Most of the rivers that are tributary to the lake of Aral are diminished by canals that carry off water for irrigation; for that reason the Oxus never reaches the lake. Besides, the Russian rivers yield less water than formerly, from the progress of culture.

The absence of lakes in the Himalaya is one of the peculiarities of these mountains. The lake of Ular, in the valley of Cashmere, is the only one of any magnitude: it is but 40 miles in circumference, and seems to be the residue of one that had filled the whole valley at some early period. There are many great lakes, both fresh and salt, in the table-land: the annular form of Lake Palte, at the northern base of the Himalaya, is unexampled, and the height of the sacred lake of Manasa, in Great Tibet, is equally so, being 17,000 feet above the level of the sea. Tibet is full of lakes, many of which produce borax, found nowhere else but in Tuscany and the Lipari Islands. As most of the great lakes on the table-land are in the Chinese territories, strangers have not had access to them;

but the Koko-nor and Lake Lop seem to be very large; the latter is said to have a surface of 2187 square miles, and there are others not inferior to it in the north. The lakes in the Altaï are beautiful, larger, and more numerous than in any other mountain chain. They are at different elevations on the terraces by which the table-land descends to the flats of Siberia, and are, owing to geological phenomena, essentially different from those which have produced the Caspian and other steppe lakes. They seem to have been hollows formed where the axes of the different branches of the chain cross, and are most numerous and deepest in the eastern Altaï. Baikal, the largest mountain lake, supposed to owe its origin to the sinking of the ground during an earthquake, has an area of 14,800 square miles, nearly equal to the half of Scotland. It lies buried in the form of a crescent amid lofty granite mountains which constitute the edge of the table-land to the south, ending in the desert of the Great Gobi, and in the north-west they gird the shore so closely that they dip into the water in many places: 160 rivers and streams are tributary to this salt lake, which drains a country probably twice the size of Britain. The river Angara, which runs deep and strong through a crevice at its eastern end, is its principal outlet, and is supposed to carry off but a small proportion of its water. Its surface is 1793 feet above the sea-level, and the climate is as severe as it is in Europe 10° farther north, yet the lake does not freeze till the middle of December, possibly from being unfathomable with a line 600 feet long.

Two hundred and eighty years before the Christian era the large fresh-water lake of Oitz, in Japan, was formed in one night by a prodigious sinking of the ground, at the same time that one of the highest and most active volcanos in that country rose from the depths of the earth.

Very extensive lakes occur in Africa, and notwithstanding the arid soil of the southern table-land, it contains the fresh-water lake of N'yassi, one of the largest, being some hundred miles long, and though narrow in proportion, it cannot be crossed in a boat of the country in less than three days, resting at night on an island, of which there are many. It begins 200 miles north from the town of Tete, on the river Zambeze, and extends from south-east to north-west to a very great but unknown distance, and between 300 and 400 miles from the Mozambique channel. No river is known to flow out of it, but it receives the drainage of the country on the south-west. No one knows what there may be on the unexplored regions of the Ethiopian desert, but Abyssinia has the large and beautiful lake of Dembea, situate in a spacious plain, the granary of the country, and so high above the sea, that spring is perpetual, though within the tropics. There are other lakes in this great projecting promontory so full of rivers, mountains, and forests, but the low lands of Soudan, the country lying along the base of the Mountains of the Moon, in the principal region of African lakes, of which the Chad, almost the size of an inland sea, is in the very centre of the continent. Its extent, and the size of its basin, are unknown, but it receives many affluents from the Mountains of the Moon, and is itself drained by the Chadda, a principal tributary of the Niger. Other lakes of less magnitude are known to exist in these regions, and there are probably many more that are unknown. Salt-water lakes are numerous on the northern boundaries of the great lowland deserts, and many fine sheets of fresh water are found in the valleys and flat terraces of the Great and Little Atlas.

Fresh-water lakes are characteristic of the higher latitudes of both continents, but those in the old continent sink into insignificance in comparison with the number and extent of those in the new. Indeed a very large portion of North America is covered with fresh water; the five principal lakes, Superior, Huron, Michigan, Erie, and Ontario, with some of their dependants, probably cover an area of 100,000 square miles, that of Lake Superior alone 43,000, which is only 7000 square miles less than the whole of England. The American lakes contain more than half the amount of fresh water on the globe. The altitude of these lakes shows the slope of the continent: the absolute elevation of Lake Superior is 627 feet, Lake Huron is 30 feet lower, Lake Erie 32 feet lower than the Huron, and Lake Ontario is 331 feet below the level of Erie. The river Niagara, which unites these two last lakes, is 33½ miles long, and in that distance it descends 66 feet; it falls in rapids through 55 feet of that height in the last half-mile, but the upper part of its course is navigable. The height of the cascade of Niagara is 162 feet on the American side of the central island, and 1125 feet wide. On the Canadian side the fall is 149 feet high, and 2100 feet wide—the most magnificent sheet of falling water known, though many are higher. The river St. Lawrence, which drains the whole, slopes 234 feet between the bottom of the cascade and the sea. The beds of lakes Superior and Ontario are respectively 165 and 336 feet below the surface of the Atlantic, affording another instance of deep indentation in the solid matter of the globe. Some lakes are decreasing in magnitude, but the contrary seems to be the case in America; between the years 1825 and 1838 Ontario rose nearly seven feet, and according to the American engineers Lake Erie had gained several feet in the same time. Lake Huron is said to be the focus of peculiar electrical phenomena, as thunder is constantly heard in one of its bays. The lakes north of this group are innumerable, the whole country, to the Arctic Ocean, is covered with sheets of water which emit rivers and streams. Lake Winnipeg, Rein-deer Lake, Slave Lake, and some others, may be regarded as the chief members of separate groups or basins, each embracing a wide extent of country almost unknown. There are also many lakes on each side of the Rocky Mountains, and in Mexico there are six or seven lakes of considerable size, though not to be compared with those in North America; the largest is the lake of Tezcuco, on the west bank of which the city of Mexico is built.

There are various sheets of water in Central America, but only two of any magnitude, namely, Lake Izaval, out of which the Rio Dulce flows into the Gulf of Mexico, and the lake of Nicaragua, in the province of that name, about 100 miles from the sea of the Antilles.

In Central America the Andes are interrupted by plains and mere hills on the isthmus of Tehuantepec and that of Panama, on each of which there is a series of lakes and rivers, which, aided by canals, might form a water communication between the Atlantic and Pacific oceans. In the former the line proposed would connect the river Huasacula, on the Gulf of Mexico, with the bay of Tehuantepec, in the Pacific. In the isthmus of Panama, the Gulf of St. Juan would be connected, by the river of that name and the large lake of Nicaragua, with the Gulf of Costa Rica. Here the watershed is only 615 feet above the sea, and of easy excavation, and the lake, situated in an extensive plain, is deep enough for vessels of considerable size.

A range of lakes goes along the eastern base of the Andes, but the greater part of them are mere lagoons or marshes; some very large, which inundate the country to a great extent in the tropical rains. There appears to be a deep hollow in the surface of the earth at the part where Bolivia, Brazil, and Paraguay meet, in which lies the Lake Xaragas, extending on each side of the river Paraguay, but, like many South American lakes, it is not permanent, being alternately inundated and dry, or a marsh. Its inundations cover 36,000 square miles. Salt and fresh-water lakes are numerous on the plains of La Plata, and near the Andes in Patagonia, resembling in this respect those in high northern latitudes, though on a narrower scale.

In the elevated mountain-valleys and table-lands of the Andes there are many small lakes of the purest blue and green colours, intensely cold, being mostly above the line of perpetual congelation. They are generally lifeless and unfathomably deep, probably the craters of extinct volcanos. The lake of Titicaca, however, in the Bolivian Andes, has an area of 4600 square miles, and is more than 120 fathoms deep, surrounded by splendid scenery. Though 12,795 feet above the level of the Pacific, its banks are clothed with turf where they are not cultivated, and in former times were the seat of advanced civilization, to which the ruins bear testimony.

The limpid transparency of the water in lakes, especially in mountainous countries, is remarkable; minute objects are visible at the bottom, through many fathoms of water. The vivid green tints, so often observed in alpine lakes, may be produced by vegetable dyes dissolved in the water, though chemical analysis has not detected them.

Lakes, being the sources of some of the largest rivers, are of great importance for inland navigation, as well as for irrigation; while by their constant evaporation they maintain the supply of humidity in the atmosphere, so essential to vegetation, besides the embellishment a country derives from their limpid and glassy waters.

CHAPTER XX.

THE ATMOSPHERE.

THE annual supply of heat which the earth receives from the sun is always the same, and it is annually radiated into space, so that it neither accumulates in the earth nor in the atmosphere. Its distribution is very unequal, but certain it is that an excess of heat in one part of the globe is compensated by a deficiency in another; an unusually warm summer is balanced by a cold one elsewhere. Diurnal variations of heat are perceptible only at a small distance below the surface, because the earth is a bad conductor, the annual heating influence of the sun penetrates much deeper. The heat which enters the earth in summer, returns during winter; and before passing into space, tempers the cold in the higher latitudes. At the equator, where the heat is the greatest, it descends deeper than elsewhere, with a diminishing intensity; but there, and everywhere throughout the globe, there is a stratum, at the depth of from 40 to 100 feet below the surface of the ground, where the temperature never varies, and is nearly the same with the mean heat of the surface.

At least one-third of the sun's heat is absorbed by the air before reaching the earth, but the atmosphere is chiefly warmed by the radiation of the sun's heat from the earth in its return to space, which takes place most abundantly when the sky is clear and blue. It is intercepted by clouds, so that a thermometer rises in cloudy weather, and sinks when the air becomes clear and calm; even a slight mist diminishes radiation from the earth, because it returns as much heat as it receives.

The superficial temperature of the earth is great at the equator, it decreases gradually towards the poles, and is an exact mean between the two at the 45th parallel of latitude; but a multitude of causes disturb this law. It is affected chiefly by the unequal distribution of land and water, by the height above the sea, by the nature of the soil, and by vegetation, so that a line drawn on a map through all the places where the mean temperature of the earth is the same, would be very far from coinciding with the parallels of latitude, but would approximate more to them near the equator.

Every thing that lives on earth depends upon the atmosphere, the source of life and heat to animated nature. The air, being a heavy and elastic fluid, decreases in density upwards according to a determinate law, so rapidly that three-fourths of it are within four miles of the earth, and all the meteoric phenomena perceptible to us, as clouds, rain, heat, and thunder, occur in that space, though the height of the atmosphere is about 50 miles. The actual pressure of the atmosphere is about 15 pounds on every square inch, diminishing of course with the height. The density is liable to continual changes from the temperature, and the attraction of the sun and moon, which produce tides similar to those in the ocean. All these changes are responded to by variations in the height of the barometer.

The air expands and becomes lighter with heat, and contracts and becomes heavier with cold ; hence at the equator, where the sun is so powerful, the light warm air is constantly ascending to the upper regions of the atmosphere, and flowing north and south towards the poles, from whence the cold heavy air rushes along the surface of the earth to supply its place between the tropics, for the same tendency to restore equilibrium exists in the air as in other fluids. The two superficial currents are relatively deflected from their meridional directions by the rotation of the earth, so that the northern current becomes a north-east wind before arriving at the tropic of Cancer, and the southern current becomes a south-east wind before it comes to the tropic of Capricorn. At the equator they so completely neutralize each other, that far at sea a candle burns without flickering. In fact, the difference of temperature puts the air in motion, and the direction of the resulting wind at every place depends upon the difference between the rotatory motion of the wind and the rotatory motion of the earth—the whole theory of the winds depends upon these circumstances.

The trade winds and monsoons are permanent, depending on the apparent motion of the sun ; but it is evident, from theory, that there must be partial winds in all parts of the earth, occasioned by local circumstances that affect the temperature of the air; consequently the atmosphere is divided into districts, both over the sea and land, in which the winds have nearly the same vicissitudes from year to year, and the regularity is greatest towards the tropics, where the causes of disturbance are fewer. In the higher latitudes it is more difficult to discover any regularity, on account of the greater proportion of land, the difference in its radiating power, and the greater extremes of heat and cold. But even there a degree of uniformity prevails in the succession of the winds. For example, in all places where north and south winds blow alternately, a vane veers through every point of the compass in the transition, and in some places the wind makes several of these girations in the course of the year. The south-westerly winds, so prevalent in the Atlantic Ocean between the 30th and 60th degrees of north latitude, are produced by the upper current being driven down to supply the superficial current which goes towards the equator; and as it has a greater rotatory motion than the earth in these latitudes, it produces a south-westerly wind. North-westerly winds prevail in the corresponding latitudes of the southern hemisphere, from the same cause. In fact, whenever the air has a greater velocity of rotation than the surface of the earth, a wind more or less westerly is produced, and, when it has less velocity of rotation than the earth, a wind having an easterly tendency results. Thus there is a perpetual exchange between the different masses of the atmosphere, the warm air tempering the cold of the higher latitudes, and the cold air mitigating the heat of the lower; it will be shown afterwards that the aërial currents are the bearers of principles on which the life of the animal and vegetable world depends. The trade winds, being constant, are essentially connected with an equatorial permanent depression in the barometer, but the mercurial column varies in every other part of the globe with a change in the density of the air and the resulting wind; indeed, the barometer gives the surest indication of an approaching change, often warning the mariner of the gale long ere it takes place. Here it may truly be said that "coming events cast their shadows before."

Since the atmosphere is chiefly warmed by heat transmitted from the earth, the temperature of the air decreases as the height above the earth increases, so that at a very small elevation the cold becomes excessive, as on the tops of mountains. This circumstance is marked by the line of perpetual snow, which is subject to many variations, but on the mountains under the equator it has a mean height of 15,207 feet, from whence it diminishes on both sides, and at last grazes the surface at the arctic and antarctic circles.

The mean annual temperature of the air would be constant on each parallel of latitude, and would decrease regularly from the equator to the poles, were it not for the inequalities in the form and nature of the surface of the globe. But these causes of disturbance are so great that lines drawn on a map through all places having the same mean annual temperature are exceedingly irregularly, except towards the equator, where they become nearly parallel to it. As the air receives most of its warmth from the earth, radiation is the principal cause of disturbance; hence the temperature is most powerfully modified by the ocean, which occupies three times as much of the surface of the globe as the land, and is more uniform in its surface, and also in radiating power. On land, the difference in the radiating force of the mountains and table-lands from that of the plains, of the deserts from grounds covered with rich vegetation, of the wet land from the dry, is the most general cause of variation; the local causes are beyond enumeration.

Places having the same mean annual temperature often differ materially in climate; in one the winters are mild and the summers cool, whereas in others the extremes of heat and cold prevail: England is an example of the first; Quebec, Petersburg, and the arctic lands are instances of the latter. It follows, as a consequence of the same quantity of heat being received annually from the sun, and annually radiated into space, that all the climates of the earth are stable, and that the vicissitudes are merely cycles that vanish after a few years. It is possible, however, that the earth may be affected by secular changes of temperature during the progress of the solar system through space.

Moisture is evaporated in an invisible form from every part of the land and water, but in very different quantities. Seven-tenths of the atmosphere rest on the ocean, therefore the sea has the greatest influence in modifying the climates on the land and supplying the air with moisture. The evaporation is greatest between the tropics, from the excess of heat, the preponderance of the ocean, and the rankness of vegetation. The average quantity of vapour decreases from the equator to the poles, and from the lower to the higher part of the atmosphere. The absolute quantity is very partial and irregular, depending everywhere on the dryness or humidity of the surface. As the vapour ascends in the atmosphere, it maintains its invisible form till it reaches a stratum of air of lower temperature, when it is condensed into clouds, and is thence precipitated in the form of rain, hail, or snow. Its dispersion and condensation are owing to the winds, the great agents in all atmospheric changes. From friction and other causes, the currents of air in the lower parts of the atmosphere run on each other horizontally; and as they generally differ in moisture, temperature, and velocity, to them is due the formation of clouds, rain, and the generation of electricity. When two masses of air of different temperatures meet, the colder, by absorbing the heat which holds the moisture in solution, occasions the particles to coalesce and form drops of water, which fall by their gravitation; and when two strata of air of different temperatures, moving rapidly in opposite directions, come into contact, an abundant fall of rain is the consequence, and, as in tropical countries the quantity of aqueous vapour is greatest, the rain-drops are largest and the rain heaviest.

The atmosphere, when clear, is almost always positively electric. The electricity arises from evaporation and the chemical changes which are perpetually in progress all over the globe; and as they sometimes generate positive and sometimes negative electricity, they occasion great local variations in the electricity of the air, but the earth itself is always in a negative state. It has been considered by some meteorologists that clouds owe their form to the electric fluid, because, when two strata of air are of different temperatures, and move in different directions, a portion of their aqueous vapour is deposited, and the electricity evolved is taken up by the remaining vapour, which causes it to assume the form of a cloud. Electricity of each kind is probably elicited by the friction of streams of air moving rapidly in different directions, and when clouds differently charged meet a storm ensues. Hail is formed when two masses of air of very different temperatures meet suddenly; hence hail is rare in tropical countries, except near mountains. The quantity of electricity in the earth and atmosphere is very great; it is constantly varying, and performs a very important part in animal and vegetable life.

Magnetism, which pervades the whole earth, is identical with electricity, although it never comes naturally into evidence. The brilliant experiments of Dr. Faraday give a new view of the magnetic condition of the substances on the surface of the globe. He found that ten of the metals are more or less magnetic, that is to say, they possess the power of attracting either pole of a magnet, and bars of these metals freely suspended between the poles of an electric magnet assume a position in the axis or line of the magnetic force, but all other substances whatever under the same circumstances are repelled by both poles of the electric magnet, and take a position at right angles to the line or current of the magnetic force. The same effect, though less powerful, was produced by a steel horse-shoe magnet. All substances are thus either magnetic or diamagnetic, except air and the gases, which are neutral. Of the metals 10 are magnetic and 16 diamagnetic: iron and bismuth are the extremes of these two conditions of matter. The inferences drawn from these discoveries by Dr. Faraday are very important: "When we consider the magnetic condition of the earth, as a whole, without reference to its possible relation to the sun, and reflect upon the enormous amount of diamagnetic matter which forms its crust, and when we remember that magnetic curves of a certain amount of force, and universal in their presence, are passing through these matters, and keeping them constantly in that state of tension, and therefore of action, we cannot doubt but that some great purpose of utility to the system, and to us its inhabitants, is thereby fulfilled."

"It is curious to see a piece of wood, or leaf, or an apple, or a bottle of water, repelled by a magnet, or the leaf of a tree taking an equatorial position. Whether any similar effects occur among the myriads of forms which in all parts of the earth's surface are surrounded by air, and subject to the action of lines of magnetic force, is a question which only can be answered by future observations. If the sun have anything to do with the magnetism of the globe, then it is probable that part of this effect is due to the action of the light that comes to us from it, and in that view the air seems most strikingly placed round our sphere, investing it with a transparent diamagnetic, which, therefore, is permeable to his rays, and at the same time moving with great velocity across them. Such conditions seem to suggest the possibility of magnetism being thence generated." Dr. Faraday's discoveries go still farther; having magnetised and electrified a ray of light, he has added another proof of the identity of these two powers. If a ray of polarized light be transmitted through certain transparent substances placed in the line of force connecting the opposite poles of an electro-magnet, it is so affected by this power that it becomes visible or invisible according as the current is flowing or not at the moment, this influence being more complete as the ray of light is more nearly parallel to the line of magnetic force, ceasing if it is perpendicular to it. The very same effect was produced with a steel horse-shoe magnet, though more feeble in degree. Mr. Christie has proved that magnetism has an influence on light direct from the sun.*

Atmospheric air is principally a mixture of oxygen and azotic gas: of 100 parts of air, 21 are oxygen gas, the source of life and heat to the animal and vegetable kingdoms; the other 79 parts are azote, or nitrogen. Besides these chief ingredients the air contains a very small quantity of ammonia, water in an invisible state, and a tenth per cent. of carbonic acid gas. The existence of the vegetable world depends upon these constituents.

* See the 7th edition of the 'Connexion of Physical Sciences: on Polarized Light and Terrestrial Magnetism.'

CHAPTER XXI.

VEGETATION—THE NOURISHMENT AND GROWTH OF PLANTS—CLASSES—BOTANICAL DISTRICTS.

In the present state of the globe a third part of its surface only is occupied by land, and probably not more than a fourth part of that is inhabited by man, but animals and vegetables have a wider range. The greater part of the land is clothed with vegetation and inhabited by quadrupeds, the air is peopled with birds and insects, and the sea teems with living creatures and plants. These organised beings are not scattered promiscuously, but all classes of them have been originally placed in regions suited to their respective wants. Many single animals and plants are indigenous only in determinate spots, while a thousand others might have supported them as well, and to many of which they have been transported by man.

The atmosphere supplies the vegetable creation with the principal part of its food; plants extract inorganic substances from the ground, which are indispensable to bring them to maturity.

The black or brown mould, which is so abundant, is the produce of decayed vegetables. When the autumnal leaves, the spoil of the summer, fall to the ground, and their vitality is gone, they enter into combination with the oxygen of the atmosphere, and convert it into an equal volume of carbonic acid gas, which consequently exists abundantly in every good soil, and is the most important part of the food of vegetables. This process is slow, and stops as soon as the air in the soil is exhausted; but the plough, by loosening the earth, and permitting the atmosphere to enter more freely, and penetrate deeper into the ground, accelerates the decomposition of the vegetable matter, and consequently the formation of carbonic acid.

In loosening and refining the mould, the common earth-worm is the fellow-labourer with man; it eats earth, and, after extracting the nutritious part, ejects the refuse, which is the finest soil, and may be seen lying in heaps at the mouth of its burrow. So instrumental is this reptile in preparing the ground, that it is said there is not a particle of the finer vegetable mould that has not passed through the intestines of a worm; thus the most feeble of living creatures is employed by Providence to accomplish the most important ends.

The food of the vegetable creation consists of carbon, hydrogen, nitrogen, and oxygen, all of which plants obtain entirely from the atmosphere, in the form of carbonic acid gas, water, and ammonia. They imbibe these three substances, and, after having decomposed them, they give back the oxygen to the air, and consolidate the carbon, water, and nitrogen into wood, leaves, flowers, and fruit.

The vitality of plants is a chemical process, entirely due to the sun's light; it is most active in clear sunshine, feeble in the shade, and nearly suspended in the night, when plants, like animals, have rest.

The atmosphere contains only one-tenth per cent. of carbonic acid gas, yet that small quantity yields enough of carbon to form the solid mass of all the magnificent forests and herbs that clothe the face of the earth, and would soon be exhausted, were it not renewed by the breath of animals, by volcanos and mineral springs, and by combustion. The green parts of plants constantly imbibe carbonic acid in the day; they decompose it, assimilate the carbon, and return the oxygen pure to the atmosphere. As the chemical action is feeble in the shade and in gloomy weather, only a part of the carbonic acid is decomposed, and then both oxygen and carbonic acid are given out by the leaves; but during the darkness of night a chemical action of a different character takes place, and almost all the carbonic acid is returned unchanged to the atmosphere, together with the moisture that is evaporated from the leaves both night and day. Thus, plants give out pure oxygen during the day, and carbonic acid and water during the night.

Since the vivifying action of the sun brings about all these changes, a superabundance of oxygen is exhaled by the tropical vegetation in a clear unclouded sky, where the sun's rays are most energetic, and atmospheric moisture most abundant. In the middle and higher latitudes, on the contrary, under a more feeble sun, and a gloomy sky subject to rain, snow, and frequent atmospheric changes, carbonic acid is given out in greater quantity by the less vigorous vegetation. But here, as with regard to heat and moisture, equilibrium is restored by the winds: the tropical currents carry the excess of oxygen along the upper strata of the atmosphere to higher latitudes, to give breath and heat to men and animals; while the polar currents, rushing along the ground, convey the surplus carbonic acid to feed the tropical forests and jungles. Harmony exists between the animal and vegetable creations: animals consume the oxygen of the atmosphere, which is restored by the exhalation of plants, while plants consume the carbonic acid exhaled by men and animals: the existence of each is thus due to their reciprocal dependence. Few of the great cosmical phenomena have only one end to fulfil; they are the ministers of the manifold designs of Providence.

When a seed is thrown into the ground the vital principle is developed by heat and moisture, and part of the substance of the seed is formed into roots, which suck up water mixed with carbonic acid from the soil, decompose it, and consolidate the carbon. In this stage of their growth plants derive their whole sustenance from the ground. As soon, however, as the sugar and mucilage of the seed appear above the ground, in the form of leaves or shoots, they absorb and decompose the carbonic acid of the atmosphere, retain the carbon for their food, and give out the oxygen in the day, and pure carbonic acid in the night. In proportion as plants grow, they derive more of their food from the air and less from the soil, till their fruit is ripened, and then their whole nourishment is derived from the atmosphere. Trees are fed from the air, after their fruit is ripe, till their leaves fall; annuals, till they die. Air-plants derive all their food from the atmosphere. The cactus semper vivens and the sedum semper vivens, which are attached to the ground only by a point, also succulent and milky-juiced plants which grow in barren ground, are almost entirely fed from the air, and even forests sometimes grow on land destitute of carbon. It is wonderful that so small a quantity of carbonic acid as exists in the air should suffice to supply the whole vegetation of the world.

Plants absorb water from the ground by their roots, they decompose it, and the hydrogen combines in different proportions with their carbonic acid to form wood, sugar, starch, gum, vegetable oils, and acids. As the green parts of plants combine with the oxygen of the air, especially during night, when the functions of plants are torpid, it is assimilated on the return of daylight, and assists in forming oils, resins, and acids. The combination of the oxygen of the air with the leaves, and also with the blossom and fruit, during night, is quite unconnected with the vital process, as it is the same in dead plants. An acid exists in the juice of every plant, generally in combination with an alkali. It must be observed, however, that these different substances are produced at different stages in the growth; for example, starch is formed in the roots, wood, stalks, and seeds, but it is converted into sugar as the fruit ripens, and the more starch the sweeter the fruit becomes. Most of these new compounds are formed between the flowering of the plant and the ripening of the fruit, and indeed they furnish the materials for the flowers, fruit, and seeds.

Ammonia, the third organic constituent of plants, is the last residue from the decay and putrefaction of animal matter. It is volatilized, and rises into the atmosphere, where it exists as a gas, but in so small a quantity that it cannot be detected by chemical analysis; yet, as it is very soluble in water, enough is brought to the ground by rain to supply the vegetable world. Ammonia enters plants by their roots along with rain-water, and is resolved within them into its constituent elements, hydrogen and nitrogen. The hydrogen aids in forming the wood, acids, and other substances before mentioned; while the nitrogen enters into every part of the plant, and forms new compounds: it exists in the blossom and fruit before it is ripe, and in the wood as albumen; it also forms gluten, which is the nutritious part of wheat, barley, oats, and all other cerealia, as well as of esculent roots, as potatoes, beet-root, &c. Nitrogen exists abundantly in peas, beans, and pulse of every kind; quinine, morphia, and other substances, are compounds of it: in short, a plant may grow without ammonia, but it cannot produce seed or fruit; the use of animal manure is to supply plants with this essential article of their food.

Thus the decomposition and consolidation of the elementary food of plants, the formation of the green parts, the exhalation of moisture by their leaves, its absorption by their roots, and all the other circumstances of vegetable life, are owing to the illuminating power of the sun. Heat can be supplied artificially in our northern climates, but it is impossible to replace the dazzling splendour of a southern sun. His illuminating influence is displayed in a remarkable degree by the cacalia ficoides: its leaves combine with the oxygen of the atmosphere during the night, and are as sour as sorrel in the morning; as the sun rises they gradually lose their oxygen, and are tasteless by noon; and by the continued action of the light they lose more and more, till towards evening they become bitter.

The blue rays of the solar spectrum have most effect on the germination of seed; the yellow rays, which are the most luminous, on the growing plant. In spring and summer the oxygen taken in by the green leaves in the night aids in the formation of oils, acids, and the other parts that contain it; but as soon as autumn comes, the vitality or chemical action of vegetables is weakened, and the oxygen, no longer given out in the day, though still taken in during the night, becomes a minister of destruction; it changes the colour of the leaves, and consumes them when they fall. Nitrogen, so essential during the life of plants, also resumes its chemical character when they die, and by its escape hastens their decay.

Although the food which constitutes the mass of

plants is derived principally from water, and the gases of the atmosphere, fixed substances are also requisite for their growth and perfection, and these they obtain from the earth by their roots. The inorganic matters are the alkalis, phosphates, silica, sulphur, iron, and others.

It has already been mentioned that vegetable acids are found in the juices of all the families of plants. They generally are in combination with one or other of the alkaline substances, as lime, soda, potash, and magnesia, which are as essential to the existence of plants as the carbonic acid by which these acids are formed: for example, vines have potash; plants used as dyes never give vivid colours without it; all leguminous plants require it, and only grow naturally on ground that contains it. None of the corn tribe can produce perfect seeds unless they have both potash and phosphate of magnesia: nor can they or any of the grasses thrive without silica, which gives the hard coating to straw, to the beard of wheat and barley, to grass, canes, and bamboos; it is even found in solid lumps in the hollows and joints of cane, known in India by the name of tabashir. To bring the cerealia to perfection, it is indispensable that in their growth they should be supplied with carbonic acid for the stalk, silica to give it strength and firmness, and nitrogen for the grain.

Phosphoric acid is found in the ashes of all vegetables, and is essential to many. Pulse contain but little of it, and on that account are less nutritious than the cerealia. The cruciform family, as cabbages, turnips, mustard, &c., must have sulphur in addition to the substances common to the growth of all plants; each particular tribe has its own peculiarities, and requires a combination suited to it.

The ocean furnishes some of the matters found in plants; the prodigious quantity of sea-water constantly evaporated carries with it salt in a volatilized state, which, dispersed over the land by the wind, supplies the ground with salt and the other ingredients of sea-water. The inorganic matters which enter plants by their roots are carried by the sap to every part of the vegetable system. The roots imbibe all liquids presented to them indiscriminately, but they retain only the substances they require at the various stages of their growth, and throw out such parts as are useless, together with the effete or dead matter remaining after the nutriment has been extracted from it. Plants, like animals, may be poisoned, but the power they have of expelling deleterious substances by their roots generally restores them to health. The feculent matter injures the soil; besides, after a time the ground is drained of the inorganic matter requisite for any one kind of plant: hence the necessity for a change or rotation of crops.

A quantity of heat is set free and also becomes latent in the various transmutations that take place in the interior of plants; so that they, like the animal creation, have a tendency to a temperature of their own, independent of external circumstances.

The quantity of electricity requisite to resolve a grain weight of water into its elementary oxygen and hydrogen is equal to the quantity of atmospheric electricity which is active in a very powerful thunder-storm; hence some idea may be formed of the intense energy exerted by the vegetable creation in the decomposition of the vast mass of water and other matters necessary for its sustenance. But there must be a compensation in the consolidation of the vegetable food, otherwise a tremendous quantity would be in perpetual activity. Possibly some part of the atmospheric electricity may be ascribed to this cause; but there is reason to believe that electricity, excited by the power of solar light, constitutes the chemical vitality of vegetation.

The colouring matter of flowers is various, if we may judge from the effect which the solar spectrum has upon their expressed juices. The colour is very brilliant on the tops of mountains and in the Arctic lands. Possibly the diminished weight of the air may have some effect, for it can scarcely be supposed that barometrical changes should be entirely without influence on vegetation.

The perfume of flowers and leaves is owing to a volatile oil, which is often carried by the air to a great distance: in hot climates it is most powerful in the morning and evening. The odour of the humeria has been perceived at the distance of three miles from the coast of South America, a species of tetracera sends its perfume as far from the island of Cuba, and the aroma of the Spice Islands is wafted out to sea. The variety of perfumes is infinite, and shows the innumerable combinations of which a few simple substances are capable, and the extreme minuteness of the particles of matter.

In northern and mean latitudes winter is a time of complete rest to the vegetable world, and in tropical climates the vigour of vegetation is suspended during the dry, hot season, to be resumed at the return of the periodical rains. Almost all plants sleep during the night; some show it in their leaves, others in their blossom. The mimosa tribe not only close their leaves at night, but their foot-stalks droop; in a clover-field not a leaf opens till after sunrise. The common daisy is a familiar instance of a sleeping flower; it shuts up its blossom in the evening, and opens its white and crimson-tipped star, the "day's eye," to meet the early beams of the morning sun; and then also "winking mary-buds begin to ope their golden eyes." The crocus, tulip, convolvulus, and many others close their blossoms at different hours towards evening, some to open them again, others never. The condrille of the walls opens at eight in the morning and closes for ever at four in the afternoon. Some plants seem to be wide awake all night, and to give out their perfume then only, or at nightfall. Many of the jessamines are most fragrant during the twilight: the olea fragrans, the daphne adorata, and the night-stock reserve their sweetness for the midnight hour, and the night-flowering sirius turns night into day. It begins to expand its magnificent sweet-scented blossom in the twilight, it is full blown at midnight, and closes, never to open again, with the dawn of day;—these are "the bats and owls of the vegetable kingdom."

Many plants brought from warm to temperate climates have become habituated to their new situation, and flourish as if they were natives of the soil; such as have been accustomed to flower and rest at particular seasons change their habits by degrees, and adapt themselves to the seasons of the country that has adopted them. It is much more difficult to transfer Alpine plants to the plains. Whether from a change of atmospheric pressure or mean temperature, all attempts to cultivate them at a lower level generally fail: it is much easier to accustom a plant of the plains to a higher situation.

Plants are propagated by seeds, offsets, cuttings, and buds; hence they, but more especially trees, have myriads of seats of life, a congeries of vital systems acting in concert, but independently of each other, every one of which might become a new plant. In this respect the fir and pine tribe are inferior to deciduous trees which lose their leaves annually, because they are not easily propagated except by seeds. It has been remarked that all plants that are propagated by buds from a common parent stock have the same duration of life: this has been noticed particularly with regard to some species of apple-trees in England.

A certain series of transitions take place throughout the lives of plants, each part being transformed and passing into another; a law that was first observed by the illustrious poet Göthe. For example, the embryo leaves pass into common leaves, these into bracteæ, the bracteæ into sepals, the sepals into petals, which are transformed into stamens and anthers, and these again pass into ovaries with their styles and stigmas, that are to become the fruit and ultimately the seed of a new plant.

Plants are naturally divided into three classes, differing materially in organization:—The cryptogamia, whose flowers and seeds are either too minute to be easily visible, or are hidden in some part of the plant, as in fungi, mosses, ferns, and lichens, which are of the least perfect organization. Next to these are the endogenous plants, which in their growth increase from the interior, as grasses and palms. In these the fresh leaves spring from the centre, and the foot-stalks of the old leaves form the outside of the stem: plants of this class are also known as monocotyledons, because they have but one seed-lobe which forms one little leaf in their embryo state. The flowers and fruit of this class are generally referable to some law in which the number three prevails, as, for example, the petals and other parts are three in number. The exogenous plants form the third class, which is the most perfect in its organization and by much the most numerous, including the trees of the forest and most of the flowering shrubs and herbs. They increase by coatings from without, as trees, where the growth of each year forms a concentric circle of wood round the pith or centre of the stem: these are also known as dicotyledonous plants, because their seeds have two lobes, which in their embryo state appear first in two little leaves above ground, like most of the European species. The parts of the flowers and fruit of this class generally have some relation to the number five.

The three botanical classes are distributed in very different proportions in different zones: endogenous plants, such as grasses and palms, are much more rare than the exogenous class. Between the tropics there are four of the latter to one of the grass or palm tribes, in the temperate zones six to one, and in the polar regions only two to one, because mosses and lichens are most abundant in the high latitudes, where exogenous plants are comparatively rare. In the temperate zones one-sixth of the plants are annuals, omitting the cryptogamia; in the torrid zone scarcely one plant in twenty is annual, and in the polar regions only one in thirty. The number of ligneous vegetables increases on approaching the equator, yet in North America there are 120 different species of forest-trees, whereas in the same latitudes in Europe there are only 34. The social plants, grasses, heaths, furze, broom, daisies, &c., which cover large tracts, are rare between the tropics, except on the mountains and table-lands and on the llanos of equatorial America.

Equinoctial America has a more extensive and richer vegetation than any other part of the world;

Europe has not above half the number of indigenous species of plants; Asia, with its islands, has somewhat less than Europe; Australia, with its islands in the Pacific, still less; and there are fewer vegetable productions in Africa than in any part of the globe of the same extent.

Since the constitution of the atmosphere is very much the same everywhere, vegetation depends principally on the sun's light, moisture, and the mean annual temperature, and it is also in some degree regulated by the heat of summer in the temperate zones. Between the tropics, wherever rain does not fall, the soil is burnt up and is as unfruitful as that exposed to the utmost rigour of frost; but where moisture is combined with heat and light, the luxuriance of the vegetation is beyond description. The abundance and violence of the periodical rains combine with the intense light and heat to render the tropical forests and jungles almost impervious from the rankness of the vegetation. This exuberance gradually decreases with the distance from the equator; it also diminishes progressively as the height above the level of the sea increases, so that each height has a corresponding parallel of latitude where the climates and floras are similar, till the perpetual snow on the mountain-tops and its counterpart in the polar regions have a vegetation that scarcely rises above the surface of the ground. Hence in ascending the Himalaya or Andes from the luxuriant plains of the Ganges or Amazons, changes take place in the vegetation analogous to what a traveller would meet with in a journey from the equator to the poles. This law of decrease, though perfectly regular over a wide extent, is perpetually interfered with by local climate and soil. From the combination of various causes, as the distribution of land and water, their different powers of absorption and radiation, together with the form, texture, and clothing of the land, and the prevailing winds, it is found that the isothermal lines, or imaginary lines drawn through places on the surface of the globe which have the same mean annual temperature, do not correspond with the parallels of latitude. Thus in North America the climate is much colder than in the corresponding European latitudes. Quebec is in the latitude of Paris, and the country is covered with deep snow four or five months in the year, and it has occurred that a summer has passed there in which not more than sixty days have been free from frost.

In the southern hemisphere, beyond the 34th parallel, the summers are colder and the winters milder than in corresponding latitudes of the northern hemisphere. Neither does the temperature of mountains vary exactly with their height above the sea; other causes, as prevailing winds, difference of radiation, and geological structure, concur in producing irregularities which have a powerful effect on the vegetable world.

However, no similarity of existing circumstances can account for whole families of plants being confined to one particular country, or even to a very limited district, which, as far as we can judge, might have grown equally well on many others. Latitude, elevation, soil, and climate, are but secondary causes in the distribution of the vegetable kingdom, and are totally inadequate to explain why there are numerous distinct botanical districts in the continents and islands, each of which has its own vegetation, whose limits are most decided when they are separated by the ocean, mountain-chains, sandy deserts, salt-plains, or internal seas. Each of these districts is the focus of families and genera, some of which are found nowhere else, and some are common to others, but, with a very few remarkable exceptions, the species of plants in each are entirely different or representative. This does not depend upon the difference in latitude, for the vegetation of the United States of North America is totally unlike that of Europe under the same isothermal lines, and even between the tropics the greatest dissimilarity often prevails under different degrees of longitude: consequently the cause of this partial distribution of plants, and that of animals also, which is according to the same law, must be looked for in those early geological periods when the earth first began to be tenanted by the present races of organised beings.

As the land rose at different periods above the ocean, each part, as it emerged from the waves, had probably been clothed with vegetation, and peopled with animals, suited to its position with regard to the equator, and to the climate and condition of the globe then being. And as the conditions and climate were different at each succeeding geological epoch, so each portion of the land, as it rose, would be characterized by its own vegetation and animals, and thus at last there would be many centres of creation, as at this day, all differing more or less from one another, and hence alpine floras must be of older date than those in the plains. The vegetation and faunas of those lands that differed most in age and place would be most dissimilar, while the plants and animals of such as were not far removed from one another in time and place would have correlative forms or family likenesses, yet each would form a distinct province. Thus, in opposite hemispheres, and everywhere at great distances, but under like circumstances, the species are representatives of one another, rarely identical: when, however, the conditions which suit certain species are continuous, identical species are found throughout, either by original creation or by migration. The older forms may have been modified to a certain extent by the succeeding conditions of the globe, but they never could have been changed, since immutability of species is a primordial law of nature. Neither external circumstances, time, nor human art, can change one species into another, though each to a certain extent is capable of accommodating itself to a change of external circumstances, so as to produce varieties even transmissable to their offspring.

The flora of Cashmere and the higher parts of the Himalaya mountains is similar to that of southern Europe, yet the species are representative, not identical. In the plains of Tartary, where from their elevation the degree of cold is not less than in the wastes of Siberia, the vegetation of one might be mistaken for that of the other; the gooseberry, currant, willow, rhubarb, and in some places the oak, hazel, cypress, poplar, and birch, grow in both, but they are of different species. The flora near the snow-line on the lofty mountains of Europe, and lower down, has also a perfect family likeness to that in high northern latitudes. In like manner many plants on the higher parts of the Chilian Andes are similiar, and even identical, with those in Terra del Fuego; nay, the Arctic flora has a certain resemblance to that of the Antarctic regions, and even occasional identity of species. These remarkable coincidences may be accounted for by the different places having been at an early geological period at the same level above the ocean, and that they continue to retain part of their original flora after their relative positions have been changed. The tops of the Chilian Andes were probably on a level with Terra del Fuego, when both were covered with the same vegetation, and in the same manner the lofty plains of Tartary may have acquired their vegetation when they were on the level of southern Siberia.

In the many vicissitudes the surface of the globe has undergone, continents formed at one period were broken up at another into islands and detached masses by inroads of the sea and other causes. Now Professor E. Forbes has shown that some of the primary floras and faunas have spread widely from their original centres over large portions of the continents before the land was broken up into the form it now has, and thus accounts for the similarity and sometimes identity of the plants and animals of regions now separated by seas,—as, for example, islands, which generally partake of the vegetation and fauna of the continents adjacent to them. Taking for granted the original creation of specific centres of plants and animals, Professor E. Forbes has clearly proved that "the specific identity, to any extent, of the flora and fauna of one area, with those of another, depends on both areas forming, or having formed, part of the same specific centre, or on their having derived their animal and vegetable population by transmission, through migration, over continuous or closely contiguous land, aided, in the case of Alpine floras, by transportation on floating masses of ice."

By the preceding laws the limited provinces and dispersion of animal and vegetable life are explained, but the existence of single species in regions very far apart has not yet been accounted for.

Very few of the exogenous or dicotyledonous plants are common to two or more countries far apart: among the few, the samolus valerandi, a common English plant, is a native of Australia; the potentilla tridendata, not found in Europe, except on one hill in Angusshire, is common on the mountains of North America; and in the Falkland Islands there are more than 30 flowering plants identical with those in Great Britain.

There are many more instances of wide diffusion among the endogenous plants, especially grasses: the phleum alpinum of Switzerland grows without the smallest variation at the Straits of Magellan, and Mr. Bunbury met with the European quaking grasses in the interior of the country at the Cape of Good Hope; but the cellular or cryptogamous class is most widely diffused—plants not susceptible of cultivation, of little use to man, and of all others the most difficult to transport. The sticta aurata, a Cornish lichen, is a native of the Cape of Good Hope, St. Helena, the West Indian islands, and Brazil; the trichomanes brevisetum grows scarcely anywhere but in Yorkshire and Madeira; and our eminent botanist, Mr. Brown, found 38 British lichens and 28 British mosses in New Holland, yet in no two parts of the world is the vegetation more dissimilar.

Some plants are concentrated in particular spots: the cinchona, which furnishes the Peruvian bark, grows only on the Andes of Loxa and Venezuela; the cedar of Lebanon is indigenous on that celebrated mountain only; and the disa grandiflora is limited to a very small spot on the top of the table-mountain at the Cape of Good Hope; but whether these are remnants whose kindred have perished by a change of physical circumstances, or centres only beginning to spread, it is impossible to say.

CHAPTER XXII.

VEGETATION OF THE GREAT CONTINENT—OF THE ARCTIC ISLANDS—AND OF THE ARCTIC AND NORTH TEMPERATE REGIONS OF EUROPE AND ASIA.

THE southern limit of the polar flora, on the great continent, lies mostly within the Arctic circle, but stretches along the tops of the Scandinavian mountains, and reappears in the high lands of Scotland, Cumberland, and Ireland, on the summits of the Pyrenees, Alps, and other mountains in southern Europe, as well as on the table-land of eastern Asia, and on the high ridges of the Himalaya.

The great European plain to the Ural Mountains, as well as the low lands of England and Ireland, were at one period covered by a sea full of floating ice and icebergs, which made the climate much colder than it now is. At the beginning of that period the Scandinavian range, the other continental mountains, and those in Britain and Ireland, were islands of no great elevation, and were then clothed with the Arctic flora, or a representative of it, which they still retain now that they form the tops of the mountain-chains, and at that time both plants and animals were conveyed from one country to another by the floating ice. It is even probable, from the relations of the fauna and flora, that Greenland,

Iceland, and the very high European latitudes, are the residue of a great northern land which had sunk down at the close of the glacial period, for there were many vicissitudes of level during that epoch. At all events it may be presumed that the elevation of the Arctic regions of both continents, if not contemporaneous, was probably not far removed in time. Similarity of circumstances had extended throughout the whole Arctic regions, since there is a remarkable similarity and occasional identity of species of plants and animals in the high latitudes of both continents, which is continued along the tops of their mountain-chains, even in the temperate zones; and there is reason to believe that the relations between the faunas and floras of Boreal America, Asia, and Europe, must have been established towards the close of the glacial period.

The flora of Iceland approaches nearer to the British than to that of any other country, yet only one in four of the Icelandic plants are known in our islands. There are 870 species in Iceland, of which more than half are flower-bearing: this is a greater proportion than is found in Scotland, but there are only 32 of woody texture. This flora is scattered in groups according as the plants like a dry, marshy, volcanic, or marine soil. Many grow to an unnatural size close to the hot-springs; thyme grows in cracks of the basin of the Great Geyser, where every other plant is petrified; and a species of chara flourishes and bears seed in a spring hot enough to boil an egg. The Icelanders make bread from metur, a species of wild corn, and also from the bulbous root of polyganum viviparum ; their greatest delicacy is the angelica archangelica ; Iceland moss, used in medicine, is an article of commerce. There are 583 species in the Feroe islands, of which 270 are flowering plants: many thrive there that cannot bear the cold of Iceland.*

ARCTIC FLORA OF THE GREAT CONTINENT.

In the most northern parts of the Arctic lands the year is divided into one long intensely cold night and one bright and fervid day, which quickly brings to maturity the scanty vegetation. Within the limit of perpetual congelation the palmetto invalis, a very minute red or orange-coloured plant, finds nourishment on the surface of the snow, the first dawn of vegetable life: it is also found colouring large patches of snow in the Alps and Pyrenees.

Lichens are the first vegetables that appear at the limits of the snow-line, whether in high latitudes or mountain-tops, and they are the first vegetation that takes possession of volcanic lavas and new islands, where they prepare soil for plants of a higher order: they grow on rocks, stones, and trees, in fact on anything that affords them moisture. More than 2400 species are already known: no plants are more widely diffused, and none afford a more striking instance of the arbitrary location of species, as

* Trevelyan's Travels in Iceland and the Feroe Islands.

they are of so little direct use to man that they could not have been disseminated by his agency. The same kinds prevail throughout the Arctic regions, and the species common to both hemispheres are very numerous. Some lichens produce brilliant red, orange, and brown dyes; and the tripe de roche, a species of gyrophora, is a miserable substitute for food, as our intrepid countryman Sir John Franklin and his brave companions experienced in their perilous Arctic journey.

Mosses follow lichens on newly-formed soil, and they are found everywhere throughout the world in damp situations, but in greatest abundance in temperate climates: 800 species are known, of which a great part inhabit the Arctic regions, constituting a large portion of the vegetation.

In Asiatic Siberia north of the 60th parallel of latitude the ground is perpetually frozen at a very small depth below the surface: a temperature of 70° below zero of Fahrenheit is not uncommon, and in some instances the cold has been 120° below zero. Then it is fatal to animal life, especially if accompanied by wind. In some places trees grow and corn ripens even at 70° of north latitude; but in the most northern parts boundless swamps, varied by lakes both of salt and fresh water, cover wide portions of this desolate country, which is buried under snow nine or ten months in the year. As soon as the snow is melted by the returning sun, these extensive morasses are covered with coarse grass and rushes, while mosses and lichens mixed with dwarf willows clothe the plains; saline plants abound, and whole districts produce diotis ceratoides.

In Nova Zembla and other places in the far north the vegetation is so stunted that it barely covers the ground, but a much greater variety of minute plants of great beauty are crowded together there in a small space than in the Alpine regions of Europe where the same genera grow. This arises from the weakness of the vegetation; for in the Swiss Alps the same plant frequently occupies a large space, excluding every other, as the dark-blue gentian, the violet-coloured pansy, the pink and yellow stone-crops. In the remote north, on the contrary, where vitality is comparatively feeble and the seeds do not ripen, thirty different species may be seen crowded together in a brilliant mass, no one having strength to overcome the rest. In such frozen climates plants may be said to live between the air and the earth, for they scarcely rise above the soil, and their roots creep along the surface, not having power to enter it. All the woody plants, as the betula lanata, the articulated willow, andromeda tetragona, with a few berry-bearing shrubs, trail along the ground, never rising more than an inch or two above it. The salix lanata, the giant of these boreal forests, never grows more than five inches above the surface, while its stem, ten or twelve feet long, lies hidden among the moss owing shelter to its lowly neighbour.

The chief characteristic of the vegetation of the Arctic regions is the predominance of perennial and cryptogamous plants, and also of the sameness of its nature, but more to the south, where night begins to alternate with day, a difference of species appears in longitude as well as in latitude. A beautiful flora of vivid colours adorns these latitudes both in Europe and Asia during their brief but bright and ardent summer, consisting of potentillas, gentians, chickweeds, saxifrages, sedums, ranunculi, spiræas, drabas, artemisias, claytonias, and many more. Such is the power of the sun and the consequent rapidity of vegetation, that these plants spring up, blossom, ripen their seed, and die, in six weeks: in a lower latitude woody plants follow these, as berry-bearing shrubs, the glaucous kalmia, the trailing azalia, and rhododendrons. The Siberian flora differs from that in the same European latitudes by the North American genera phlox, mitella, claytonia, and the predominance of asters, solidago, spiræa, milk-vetches, wormwood and the saline plants, goosefoot, and saltworts.

Social plants abound in many parts of the northern countries, as grass, heath, furze, and broom: the steppes are an example of this on a very extensive scale. Both in Europe and Asia they are subject to a rigorous winter, with deep snow and chilling blasts of wind; and as the soil generally consists of a coating of vegetable mould over clay, no plants with deep roots thrive upon them; hence the steppes are destitute of trees, and even bushes are rare except in ravines: the grass is thin, but nourishing. Hyacinths and some other bulbs, mignionette, asparagus, liquorice, and wormwood, grow in the European steppes; the two last are peculiarly characteristic. The nymphæa nilumba grows in one spot five miles from the town of Astracan, and nowhere else in the wide domains of Russia: the leaves of this beautiful aquatic plant are often two feet broad, and its rose-coloured blossoms are very fragrant. It is also native in India and Tibet, where it is held sacred, as it was formerly in Egypt, where it is said to be extinct: it is one of the many instances of a plant growing in countries far apart.

Each steppe in Siberia has its own peculiar plants: the peplis and camphorasina are peculiar to the steppe of the Irtish, and the amaryllus tartarica abounds in the meadows of eastern Siberia, where the vegetation bears a great analogy to that of north-western America: several genera and species are common to both.

Half the plants found by Wormskiold in Kamtchatka are European, with the exception of eight or ten, which are American. Few European trees grow in Asiatic Siberia, notwithstanding the similarity of climate, and most of them disappear towards the rivers Tobol and Irtish.

In Lapland and in the high latitudes of Russia large tracts are covered with birch-trees, but the pine and fir tribe are the principal inhabitants of the north. Prodigious forests of these are spread over the mountains of Norway and Sweden, and in European Russia 200,000,000 acres are clothed with these coniferæ alone, or occasionally mixed with willows, poplars, and alders. Although soils of pure sand and lime are absolutely barren, yet they generally contain enough of alkali to supply the wants of the fir and pine tribes, which require ten times less than oaks and other deciduous trees.

The Siberian steppes are bounded on the south by great forests of pine, birch, and willow: poplars, elms, and Tartarian maple overhang the upper courses of the noble rivers which flow from the mountains to the Frozen Ocean, and on the banks of the Yenessei the pinus cimbra, or Siberian pine, with edible fruit, grows 120 feet high. The Altaï are covered nearly to their summit with similar forests, but on their greatest heights the stunted larch crawls on the ground, and the flora is like that of northern Siberia: round the lake Baikal the pinus cimbra grows nearly to the snow-line.

Forests of black birch are peculiar to Da-Ouria, where there are also apricot and apple trees, and rhododendrons, of which a species grows in thickets on the hills, with yellow blossoms. Here and everywhere else throughout this country are found all the species of caragana, a genus entirely Siberian. Each terrace of the mountains and each steppe on the plains has its peculiar plants, as well as some common to all: perennial plants are more numerous than annuals.

If temperature and climate depended upon latitude alone, all Asia between the 50th and 30th parallels would have a mild climate; but that is far from being the case, on account of the structure of the continent, which consists of the highest table-lands and the lowest plains on the globe.

The table-land of Tibet, where it is not cultivated, has the character of great sterility, and the climate is as unpropitious as the soil: frost, snow, and sleet begin early in September, and continue with little interruption till May; snow, indeed, falls every month in the year. The air is always dry, because in winter moisture falls in the form of snow, and in summer it is quickly evaporated by the intense heat of the sun. The thermometer sometimes rises to 144° of Fahrenheit in the sun, and even in winter his direct rays have great power for an hour or two, so that a variation of 100° in the temperature of the air has occurred in twelve hours. Notwithstanding these disadvantages there are sheltered spots which produce most of the European grain and fruits, though the natural vegetation bears the Siberian character, but the species are quite distinct. The most common indigenous plants are Tartarian furze and various prickly shrubs resembling it, gooseberries, currants, hyssop, dog-rose, dwarf sow-thistle, equisetum, rhubarb, lucern, and asafœtida, on which the flocks feed. Prangos, an umbelliferous plant with broad leaves and scented blossom, is peculiar to Ladak and Tibet. Mr. Moorcroft says it is so nutritious, that sheep fed on it become fat in twenty days. There are three species of wheat, three of barley, and two of buckwheat, natives of the lofty table-land, where the sarsinh is the only fruit known to be indigenous. Owing to the rudeness of the climate trees are not numerous, yet on the lower declivities of some mountains there are aspens, birch, yew, ash, Tartaric oak, various pines, and the pavia, a species of horse-chesnut. Much of the table-land of Tartary is occupied by the Great Gobi and other deserts of sand, with grassy steppes near the mountains; but of the flora of these regions we know nothing.

FLORA OF BRITAIN AND OF MIDDLE AND SOUTHERN EUROPE.

The British Islands afford an excellent illustration of distinct provinces of animals and plants, and also of their migration from other centres. Professor E. Forbes has determined five botanical districts, four of which are restricted to limited provinces, whilst the fifth, which comprehends the great mass of British plants, is everywhere, either alone or mixed with the others. The first includes the flora of the mountain districts in the west and south-west of Ireland, which is similar to that in the north of Spain. The flora in the south of England and south-east of Ireland is different from that in all other parts of the British Islands, but is intimately related to that of the Channel Islands and the French coast opposite to them.

In the south-east of England the flora is like that on the adjacent coast of France. The tops of the Scottish mountains are the focus of a separate flora, a few of whose plants are found also on the summits of the mountains in Cumberland and Wales, and Scandinavian plants are mingled with it in Scotland. The fifth, of more recent origin than the Alpine flora, includes all the ordinary flowering plants, as the common daisy and primrose, hairy-ladies' smock, upright meadow crowfoot, and the lesser celandine, together with our common trees and shrubs, has migrated from Germany before England was separated from the continent of Europe by the British Channel. It can be distinctly traced in its progress across the island, but the migration was not completed till after Ireland was separated from England by the Irish Channel, and that is the reason why many of the ordinary English plants, animals, and reptiles are not found in the sister island, for the migration of animals was simultaneous with that of plants, and took place between the last of the tertiary periods and the historical epoch, that of man's creation: it was extended also over a great part of the Continent.

Deciduous trees are the chief characteristic of the temperate zone of the old continent, more especially of middle Europe: these thrive best in soil produced by the decay of the primary and ancient volcanic rocks, which furnish abundance of alkali. Oaks, elms, beech, ash, larch, maple, lime, alder, and sycamore, all of which lose their leaves in winter, are the prevailing vegetation, occasionally mixed with fir and pine.

The undergrowth consists of wild apple, cherry, yew, holly, hawthorn, broom, furze, wild rose, honeysuckle, clematis, &c. The most numerous and characteristic herbaceous plants are the umbelliferous class, as carrot and anise, the campanulas, the chicoraceæ, a family to which lettuce, endive, dandelion, and sow-thistle belong. The cruciform tribe, as wallflower, stock, turnip, cabbage, cress, &c., are so numerous, that they form a distinguishing feature in the botany of middle Europe, to which 45 species of them belong. This family is almost confined to the northern hemisphere, for of 800 known species, only 100 belong to the southern, the soil of which must contain less sulphur, which is indispensable for these plants.

In the Pyrenees, Alps, and other high lands in Europe, the gradation of botanical forms from the summit to the foot of the mountains is similar to that which takes places from the Arctic to the middle latitudes of Europe. The analogy, however, is true only when viewed generally, for many local circumstances of climate and vegetation interpose, and although the similarity of botanical forms is very great between certain zones of altitude and parallels of latitude, the species are for the most part different.

Evergreen trees and shrubs become more frequent in the southern countries of Europe, where about a fourth part of the ligneous vegetation never entirely lose their leaves. The flora consists chiefly of ilex, oak, cypress, hornbeam, sweet chesnut, laurel, laurestina, the apple tribe, manna, and the flowering ash, carob, jujub, juniper, terebinthinas, and lentiscus pistaccio, which yield resin and mastick, arbutus, myrtle, jessamine, yellow and white, various pines, as the pinus maritima, and pinus pinea, or stone pine, which forms so picturesque a feature in the landscape of southern Europe. The most prevalent herbaceous plants are caryophyllæ, as pinks, stellaria and arenarias, and also the labiate tribe, mint, thyme, rosemary, lavender, with many others, all remarkable for their aromatic properties, and their love of dry situations. Many of the choicest plants and flowers, which adorn the gardens and grounds in northern Europe, are indigenous in these warmer countries; the anemone, tulip, mignionette, narcissus, gladiolus, iris, asphodel, amaryllis, carnation, &c. In Spain, Portugal, Sicily, and the other European shores of the Mediterranean, tropical families begin to appear in the arums, plants yielding balsams, oleander, date and palmetto palms, and grasses of the order panicum or millet, cyperaceæ, or sedges, aloe and cactus. In this zone of transition there are six herbaceous for one woody plant.

FLORA OF TEMPERATE ASIA.

The vegetation of western Asia approaches nearly to that of India at one extremity, and Europe at the other; of 281 genera of plants which grow in Asia Minor and Persia, 109 are European. Syria and Asia Minor form a region of transition, like the other countries on the Mediterranean, where the plants of the temperate and tropical zones are united.

We owe many of our best fruits and sweetest flowers to these regions. The cherry, almond, oleander, syringa, locust tree, &c., come from Asia Minor; the walnut, peach, melon, cucumber, hyacinth, ranunculus, come from Persia; the date palm, fig, olive, mulberry, and damask rose, come from Syria; the vine and apricot are Armenian, the latter grows also everywhere in middle and northern Asia. The tropical forms met with in more sheltered places are the sugar-cane, date and palmetto palms, mimosas, acacias, asclepea gigantea, and other arborescent apocineæ. On the mountains south of the Black Sea, American types appear in rhododendrons, and the azalea pontica, and herbaceous plants are numerous and brilliant in these countries.

The table-land of Persia, though not so high as that of eastern Asia, resembles it in the quality of the soil, which is chiefly clayey, sandy, or saline, and the climate is very dry; hence vegetation is poor, and consists of thorny bushes, acacias, mimosas, tamarisk, jujub, and asafœtida. Forests of oak cover the Lusistan mountains, but the date palm is the only produce of the parched shores of the Arabian Gulf and of the oases on the Persian table-land. In the valleys, which are beautiful, there are clumps of Oriental plane and other trees, hawthorn, tree roses, and many of the odoriferous shrubs of Arabia Felix.

Afghanistan produces the seedless pomegranate. The true indigo grows in the lower offsets of the Hindoo-coosh, where the valleys are covered with clover, thyme, violets, and many odoriferous plants: the greater part of the trees in the mountains are of European genera, though all the species of plants, both woody and herbaceous, are peculiar.

Hot arid deserts bound India on the west, where the stunted and scorched vegetation consists of tamarisks, thorny acacia, deformed euphorbiæ, and almost leafless thorny trees, shaggy with long hair, by which they imbibe moisture and carbon from the atmosphere. Indian forms appear near Delhi, in the genera flacourtia and others, mixed with Syrian plants. East of this transition the vegetation becomes entirely Indian, except on the higher parts of the mountains, where European types prevail.

The Himalaya Mountains form a distinct botanical district. Immediately below the snow-line the flora is almost the same with that on the high plains of Tartary, to which may be added rhododendrons and andromedas, and among the herbaceous plants a primrose appears. Lower down vast tracts are covered with prostrate bamboos, and European forms become universal, though the species are Indian, as gentians, plantagos, campanulas, and gale. There are extensive forests of coniferæ, consisting chiefly of pinus excelsa, deodora, and morinda, with many deciduous forest and fruit-trees of European genera. A transition from this flora to a tropical vegetation takes place between the altitudes of 9000 and 5000 feet, because the rains of the monsoons begin to be felt in this region, which unites the plants of both. Here the scarlet and other rhododendrons grow luxuriantly; walnuts, and at least ten species of oak, attain a great size, one of which, the quercus semicarpifolia, has a clean trunk from 80 to 100 feet high. Geraniums, labiate plants, are mixed in sheltered spots with the tropical genera of scitamineæ, or the ginger tribe; bignonias and balsams, and camelias, grow on the lower part of this region.

It is remarkable that Indian, European, American, and Chinese forms are united in this zone of transition, though the distinctness of species still obtains: the triosteum, a genus of the honeysuckle tribe, is American; the abelia, another genus of the same, together with the camelia and tricyrtis, are peculiarly Chinese; the daisy and wild thyme are European. A few of the trees and plants mentioned descend below the altitude of 5000 feet, but they soon disappear on the hot declivities of the mountain, where the erythrina, monosperma, and bombex heptaphyllum, are the most common trees, together with the millingtoneæ, a tribe of large timber trees, met with everywhere between the Himalaya and 10° north latitude. The shorea robusta, dalbergia, and cedrela, a genus allied to mahogany, are the most common trees in the forests of the lower regions of these mountains.

The temperate regions of eastern Asia, including Chinese Tartary, China, and Japan, have a vegetation totally different from that of any other part of the globe similarly situated, and shows in a strong point of view the distinct character which vegetation assumes in different longitudes. In Mandshuria and the vast mountain-chains that slope from the eastern extremity of the high Tartarian table-land to the fertile plains in China, the forests and flora are generally of European genera, but Asiatic species; in these countries the buck-thorn and honeysuckle tribes are so numerous as to give a peculiar character to the vegetation.

The transition zone in this country lies between the 35th and 27th parallels of north latitude, in which the tropical flora is mixed with that of the northern provinces. The prevailing plants on the Chinese low grounds are glycyne, hortensia, the camphor laurel, stillingia sebifera, or wax tree, clerodendron, hibiscus sinensis, thuia orientalis, olea fragrans, the sweet blossoms of which are mixed with the finer teas to give them flavour; melia azedarach, or Indian pride, the paper mulberry, and others of the genus, and camelia sasanqua, which covers hills in the province of Kiong-si. The tea-plant, and other species of camelia, grow in many parts; the finest tea is the produce of a low range of hills between the 30th and 32nd parallels, an offset from the great chain of Peling. The tea-plant is not confined to China, it grows on the mountains of Assam, and as some species of the camelia tribe are indigenous in the temperate regions of the Himalaya, it might probably be cultivated in that range.

The climate of Japan is milder than its latitude would indicate, owing to the influence of the surrounding ocean. European forms prevail in the high lands, as they do generally throughout the mountains of Asia and the Indian Archipelago, with the difference of species, as abies cembra, strobus, and larix. The Japanese flora is similar to the Chinese, and there are 30 American plants, besides others of Indian and tropical climates. These islands, nevertheless, have their own peculiar flora, distinct in its nature; as the saphora, corchorus, aukuba, mespilus, and pyrus Japonica, rhus vernix, oralis cordata, the anise tree, daphne odorata, the soap tree, various species of the calecanthus tribe, the custard apple, the khair mimosa, which yields the catechu, the leechee, the sweet orange, the cycas revoluta, a plant resembling a dwarf palm, with various other fruits. Many tropical plants mingle with the vegetation of the cocoa-nut and fan-palms.

Thus the vegetation in Japan and China is widely different from that in the countries bordering the Mediterranean, though between the same parallels of latitude. In the tropical regions of Asia, where heat and moisture are excessive, the influence of latitude vanishes altogether, and the peculiarities of the vegetation in different longitudes become more evident.

CHAPTER XXIII.

FLORA OF TROPICAL ASIA — OF THE INDIAN ARCHIPELAGO, INDIA, AND ARABIA.

Tropical Asia is divided by nature into three distinct botanical regions: the Malayan peninsula, with the Indian Archipelago: India, south of the Himalaya, with the island of Ceylon; and the Arabian peninsula. The two first have strong points of resemblance, though their floras are peculiar.

FLORA OF THE INDO-CHINESE PENINSULA AND THE INDIAN ARCHIPELAGO.

Many of the vegetable productions of the peninsula beyond the Ganges are the same with those of India, mixed with the plants of the Indian Archipelago, so that this country is a region of transition, though it has a splendid vegetation of innumerable native productions, dyes of the most vivid hues, spices, medicinal plants, and many with the sweetest perfume. The soil in many places yields three crops in the year; the fruits of India, and most of those of China, come to perfection in the low lands. The areng palm is peculiarly characteristic of the Malayan peninsula; it yields sago and wine, is an ugly plant, covered with black fibres like coarse horsehair, so strong that cordage is made of it. Teak is plentiful; almost all that is used in Bengal comes from the Birman empire, though it is less durable than that of the Malabar coast. The hopœa odorata is so large that a canoe is made of a single trunk; the cardonia integrifolia is held in such veneration that every Birman house has a beam of it.

There are seven species of native oak in the forests; the mimosa catechu, which furnishes the terra japonica used in medicine; the trees which produce varnish and stick-lac; the glyphyrea nitida, a myrtle, the leaves of which are used as tea in Bencoolen, called by the natives the tree of long life. The coasts are wooded by the heriliera robusta, a large tree which thrives within reach of the tide; bamboos with stems a foot and a half in diameter grow in dense thickets in the low lands. The palmyra palm and the borassus flabelliformis grow in extensive groves in the valley of the Irawaddie: it is a magnificent tree, often 100 feet high, remarkable for its gigantic leaves, one of which would shelter twelve men.

The anomalous trees the zamas and cycades, somewhat like a palm with large pinnated leaves, but of a different genus, are found here and in tropical India; those in America are of a different species. Orchideæ and tree-ferns are innumerable in the woody districts of the peninsula.

The vegetation of the Indian Archipelago is gorgeous beyond description; although in many instances it bears a strong analogy to that of the Malayan peninsula, tropical India, and Ceylon, still it is in an eminent degree peculiar. The height of the mountains causes variety in the temperature sufficient to admit of the growth of dammer pines, oaks, rhododendrons, magnolias, valerians, honey-suckles, bilberries, gentians, oleasters, and other European orders of woody and herbaceous plants; yet there is not one species in common.

Jungle and dense pestilential woods entirely cover the smaller islands and the plains of the larger; the coasts are lined with thickets of mangroves, a matted vegetation of forest trees, palms, bamboos, and coarse grass, entwined with climbing and creeping plants, and overgrown by orchideous parasites in myriads. The forest trees of the Indian Archipelago are almost unknown; teak and many of the continental trees grow there, but the greater number are peculiarly their own. The naturalist Rumphius had a cabinet inlaid with 400 kinds of wood, the produce of Amboyna and the Molucca Islands.

Sumatra, Java, and the adjacent islands, are the region of the caryota urens and of the dryobalanops camphora of the laurel tribe, in the stems of which solid lumps of camphor are found. All the trees of that order, and of several others, are peculiar to these islands, and 78 species of trees and shrubs of the melastoma tribe grow there and in continental India. There are thickets of the sword-leaved vaquois tree and of the pandanus or screw-pine, a plant resembling the anana, with a blossom like that of a bulrush, very odoriferous, and in some species edible.

This is the region of spices, which are very limited in their distribution: the myristica moschata, the nutmeg, and mace-plant, is confined to the Banda Islands, but it is said to have been discovered lately in New Guinea. The Amboyna group is the focus of the caryophyllus aromaticus, a myrtle, the buds of which are known as cloves. Various species of cinnamon and cassia, both of the laurel tribe, together with varieties of pepper, different from those in India and Ceylon, grow in this archipelago. Some of the most excellent fruits are indigenous here only, as the dourio, the ayer ayer, langsat, the choapa of Molucca, peculiar kinds of orange, lemon, and citron, with others known only by name elsewhere. Those common to the continent of India are the jamrose, rose-apple, jack, various species of bread-fruit, mango, mangosteen, and the banana, which is luxuriant.

Here the nettle tribe assume the most pernicious character, as the upas tree of Java, one of the most deadly vegetable poisons, and even the plants resembling our common nettle are so acrid that the sting of one in Java occasions not only pain but illness, which lasts for days. A nettle in the island of Timor, called by the natives the "Devil's leaf," is so poisonous that it produces long illness and even death. The chelik, a shrub growing in the dense forests, produces a poison even more deadly than the upas. Some of the fig genus, which belongs also to the natural order of nettles, have acrid juices. Trees of the cashew tribe have a milky sap: the fine japan lacquer is made from the juice of the stagmaria verniciflua. Palms are very splendid here, generally of peculiar species and limited in their distribution, as the nissa and Barringtonia. No country is richer in club-mosses and orchideous plants, which overrun the trees in thousands in the deep dark mountain-forests, choked by huge creeping plants, an undergrowth of gigantic grasses, through which not a ray of light penetrates.

Sir Stamford Raffles describes the vegetation of Java as "fearful." In these forests the air is heavy, charged with dank and deadly vapours, never agitated by a breath of wind; the soil, of the deepest black vegetable mould, always moist and clammy, stimulated by the fervid heat of a tropical sun, produces trees whose stems are of a spongy texture from their rapid growth, loaded with parasites, particularly the orchideous tribe, of which no less than 300 species are peculiar to that island. Tree-ferns are in the proportion of one to twenty of the other plants, and form a large portion of the vegetation of Java and all these islands; and there are above 200 tropical species of club-mosses growing to the height of three feet, whereas in cold countries they creep on the ground.

The Rafflesia, of which there are four genera, are the most singular productions of this archipelago. The most extraordinary is common to Java and Sumatra, where it was discovered by Mr. Arnold, and therefore is called Rafflesia Arnoldi. It is a parasitical plant, with buds the size of an ordinary cabbage, and the flower, which smells of carrion, is of a brick-red colour, three feet and a half in diameter: that found by Mr. Arnold weighed fifteen pounds, and the cup in its centre could contain twelve pints of liquid.

According to Sir Stamford Raffles there are six distinct climates in Java, from the top of the mountains to the sea, each having an extensive indigenous vegetation. No other country can show an equal abundance and variety of native fruit and esculent vegetables. There are 100 varieties of rice, and of fragrant flowers, shrubs, and ornamental trees the number is infinite. Abundant as the orchideæ are in Java, Ceylon, and the Burmese empire, these countries possess very few that are common to them all, so local is their distribution. Ferns are more plentiful in this archipelago than elsewhere: tree-ferns are found chiefly between or near the tropics, in airless damp places.

INDIAN FLORA.

The plains of Hindostan are so completely sheltered from the Siberian blasts by the high table-lands of Tartary and the Himalaya Mountains, that the vegetation at the foot of that range already assumes a tropical character. In the jungles and lower ridges of the fertile valley of Nepal, and on the dark and airless recesses of the Silhet forests, arborescent ferns and orchideous plants are found in profusion, scarcely surpassed even in the islands of the Indian Archipelago—indeed the marshy

Tariyane is full of them. The lowest ranges of the Himalaya, the pestilential swamp of the Tariyane, the alluvial ridges of the hills that bound it on the south, and many parts of the plains of the Ganges, are covered with primeval forests, which produce whole orders of large timber trees, frequently overrun with parasitical loranthe.

The native fruits of India are many; the orange tribe is almost all of Indian origin, though some of the species are now widely spread over the warmer parts of the other continents and the more distant countries of Asia. Two or three species are peculiar to Madagascar; one is found in the forests of the Essequibo and another in Brazil, which are the only exceptions known. The limonia laureola grows on the tops of the high Asiatic mountains, which are covered with snow several months in the year; and the wampee, a fruit much esteemed in China and the Indian Archipelago, is produced by a species of this order. The vine grows wild in the forests; plantain, banana, jamrose, guava, mango, mangosteen, date, areca, palmyra, cocoa-nut, and gameto-palms are all Indian, also the gourd family. The Scitaminæ, or ginger-tribe, are so numerous, that they form a distinguishing and beautiful feature of Indian botany: they produce ginger, cardamoms, and turmeric. The flowers peculiar to India are brilliant in colours, but generally without odour, except the rose and some jessamines.

The greater part of the trees and plants mentioned belong also to tropical India, where vegetation is still more luxuriant; a large portion of that magnificent country, containing 1,000,000 square miles, has been cultivated time immemorial, although vast tracts still remain in a state of nature. Those extensive mountain-chains which traverse and surround the Deccan are rich in primeval forests of stupendous growth with dense underwood. The most remarkable of these trees are the Indian cotton-tree and the dombeya, which is of the same order; that which produces the Trincomalee wood, used for building boats at Madras; the red-wood tree, peculiar to the Coromandel coast, the satin-wood, the superb butea frondosa, the agallshium tribe, which yields the odorous wood of aloes mentioned in Scripture, the melaleuca leucadendron and the melaleuca cujapute, from which the oil is prepared. The dragon-tree is a native of India, though not exclusively, as some of the best specimens grow in the Azores and Madagascar, where it is planted for hedges. Sanders-wood and dragon's-blood are obtained from the pterocarpus sandalinus and draco; the sappan-tree gives a purple dye: these are all of the leguminous or bean tribe, of which there are 452 Indian species: ebony grows in these tropical regions, in Mauritius, and the south coast of Africa.

Some of the fig tribe are among the most remarkable vegetable productions of India for gigantic size and peculiarity of form, which renders them valuable in a hot climate from the shade which their broad-spreading tops afford. Some throw off shoots from their branches, which take root on reaching the ground, and after increasing in girth with wonderful rapidity, produce branches which also descend to form new roots, and this process is continued till a forest is formed round the parent tree. Mr. Reinwardt saw in the island of Simao a large wood of the ficus Benjamina which sprung from one stem. The ficus Indicus, or banyan tree, is another instance of this wide-spreading growth; it is found in the islands, but is in greatest perfection around the villages in the Cirçar mountains. The camphor genus is mostly Indian, as well as many more of the laurel tribe of great size. The banana is the most generally useful tree in this country; its fruit is food, its leaves are applied to many domestic purposes, and flax fit for making muslin is obtained from its stem.

Palms, the most stately and graceful of the vegetable productions of tropical regions, are abundant in India, in forests, in groups, and in single trees. Some species grow at the limit of perpetual snow, some 900 feet above the sea, others in valleys and on the shores of the continent and islands. They decrease in number and variety as the latitude increases, and terminate at Nice, in 44° N. lat., their limit in the great continent. The leaves of some are of gigantic size, and all are beautiful, varying in height from the slender calamus rudentum, 500 feet high, to the chamærops humilis, not more than 15 or 20. Different species yield wine, oil, wax, flour, sugar, thread, and rope; weapons and utensils are made of their stems and leaves; they serve for the construction of houses; the cocoa-nut palm gives food and drink; sago is made from all except the areca catechu, the fruit of which, the betel-nut, is used by the natives for its intoxicating quality. A few of the species are widely spread, for example the cocoa-nut palm, though they are in general very limited in their distribution.

The island of Ceylon, which may be regarded as the southernmost extremity of the Indian peninsula, is very mountainous, and rivals the islands of the Indian Archipelago in luxuriance of vegetable productions, and in some respects bears a strong resemblance to them. The laurel, the bark of which is cinnamon, is indigenous, and one of the principal sources of the revenue of Ceylon. The taleput leaves of the areca palm are of such enormous size, that they are applied to many uses by the Cingalese: in ancient times stripes of the leaf were written upon with a sharp style, and served as books. The sandal-wood of Ceylon is of a different species from that of the South Sea Islands, and its perfume more esteemed. Indigo is indigenous, and so is the choya, whose roots give a scarlet dye. The mountains produce a great variety of beautiful woods used in cabinet-work. It is a remarkable circumstance in the distribution of plants, that the orchideæ are very numerous in this island, and that there should be none in the Indian peninsula.

ARABIAN VEGETATION.

The third division of the tropical flora of Asia is the Arabian, which differs widely from the other two, and is chiefly marked by trees yielding balsams. Oceans of barren sand extend to the south, from Syria through the greater part of Arabia, varied only by occasional oases in those spots where a spring of water has reached the surface; there the prevalent vegetation consists of the grasses, holcus and panicum dicotomum growing under the shade of the date-palm; mimosas and stunted prickly bushes appear here and there in the sand. There is verdure on the mountains, and along some of the coasts, especially in the province of Yemen, which has a flora of its own, and is the native country of coffea, which is now cultivated over half the globe. Most of the coffea used is the progeny of a single plant brought from Mocha to the botanic garden at Amsterdam, by Van Hoorn, the governor of Batavia, in the year 1718. Plants were sent to Surinam, from whence they spread rapidly over the warm parts of America and the West Indian Islands. The keura odorifera, a superb tree, with agreeable perfume, eight species of figs, the three species of amyris gileadensis, or balm of Gilead, opobalsamum also yielding balsam, and the kataf, from which myrrh is supposed to come, are peculiar to Arabia. Frankincense is said to be the produce of the Boswellia serrata; and there are many species of acacia, among others the acacia arabica, which produces gum arabic. The arak and tamarind trees connect the botany of Arabia with that of the West Indies, while it is connected with that of the Cape of Good Hope by Stapelias, mesembryanthemums, and liliaceous flowers. The character of Arabian vegetation, like that of other dry hot climates, consists in its odoriferous plants and flowers.

CHAPTER XXIV.

AFRICAN FLORA—FLORA OF AUSTRALIA, NEW ZEALAND, NORFOLK ISLAND, AND OF POLYNESIA.

The northern coast of Africa, and the range of the Atlas generally, may be regarded as a zone of transition, where the plants of southern Europe are mingled with those peculiar to the country; half the plants of northern Africa are also found in the other countries on the shores of the Mediterranean. Of 60 trees and 248 shrubs which grow there, 100 only are peculiar to Africa, and about 18 of these belong to its tropical flora. There are about six times as many herbaceous plants as there are trees and shrubs; and in the Atlas Mountains, as in other chains, the perennial plants are much more numerous than annuals. Evergreens predominate, and are the same as those on the other shores of the Mediterranean. The pomegranate, the locust tree, the oleander, and the palmetto abound; and the cistus tribe give a distinct character to the flora. The sandarach, or thuia articulata, peculiar to the northern side of the Atlas Mountains and to Cyrenaica, yields close-grained hard timber, used for the ceiling of mosques, and is supposed to be the shittim-wood of Scripture. The Atlas produces seven or eight species of oak, various pines, especially the pinus maritima, and forests of the Aleppo pine in Algiers. The sweet-scented arborescent heath and erica scoparia are native here, also in the Canary Islands and the Azores, where the tribe of house-leeks characterises the botany. There are 534 phanerogamous plants, or such as have the parts of fructification evident, in the Canary Islands; the pinus canarienses is peculiar, and also the dracœnæ, which grow in perfection here. The stem of the dracœna draco, of the Villa Oratavas in Teneriffe, measures 46 feet in circumference at the base of the tree, which is 75 feet high. It is known to have been an object of great antiquity in the year 1402, and is still alive, bearing blossoms and fruit. If it be not an instance of the partial location of plants, there must have been intercourse between India and the Canary Islands in very ancient times.

Plants with bluish-green succulent leaves are characteristic of tropical Africa and its islands; and though the group of the Canaries has plants in common with Spain, Portugal, Africa, and the Azores, yet there are many species, and even genera, which are found in them only; and the height of the mountains causes much variety in the vegetation.

On the continent, south of the Atlas, a great change of soil and climate takes place; the drought on the borders of the desert is so excessive that no trees can resist it, rain hardly ever falls, and the scorching blasts from the south speedily dry up any moisture that may exist; yet in consequence of what descends from the mountains, the date-palm forms large forests along their base, which supply the inhabitants with food, and give shelter to crops which could not otherwise grow. Stunted plants are the only produce of the desert, yet large tracts are covered with the pennisetum dichotomum, a harsh prickly grass, which, together with the alhagi maurosin, are the food of camels.

The plants peculiar to Egypt are acacias, mimosas, cassias, tamarisks, the lotus nymphæa, the blue lotus, the papyrus, from which probably the first substance used for writing upon was made, and has left its name to that we now use: also the ziziphus or jujub, various mesembryanthemums, and most of the plants of Barbary grow here. The date-palm is not found higher on the Nile than Thebes, where it gives place to the doom-palm, or crucifera Thebaica, peculiar to this district, and singular as being the only palm that has a branched stem.

The eastern side of equatorial Africa is less known than the western, but the floras of the two countries, under the same latitude, have little affinity; on the eastern side the rubiaceæ, the euphorbiæ, a race peculiarly African, and the malviaceæ, are most frequent. The genera danais of the coffea tribe distinguish the vegetation of Abyssinia, also the dombeya, the senaceæ, a species of vine, various jessamines, a beautiful species of honeysuckle; and Bruce says the caper-tree grows to the height of the elm, with white blossoms, and fruit as large as a peach. The daroo, or ficus sycamoris, and the arak tree, are native. The kollquall, or euphorbia antiquorum, grows 40 feet high on the plain of Baharnagach, in the form of an elegant branched candelabrum, covered with scented fruit. The kantuffa, or thorny shrub, is so great a nuisance from its spines, that even animals avoid it. The erythrina Abyssinica bears a poisonous red bean with a black spot, used by the Shangalla and other tribes for ages, as a weight for gold, and by the women as necklaces. Mr. Rochet has lately brought some seeds of new grain from Shoa, that are likely to be a valuable addition to European cerealia.

The vegetation of tropical Africa, on the west, is known only along the coast, where some affinity with that of India may be observed. It consists of 573 species of flower-bearing plants, and is distinguished by a remarkable uniformity, not only in orders and genera, but even in species, from the 16° of north latitude to the river Congo in 6° of south latitude. The most prevalent are the grasses and bean tribes, the cyperaceæ, rubiaceæ, and the compositæ. The Adamsonia, or boabab of Senegal, is one of the most extraordinary vegetable productions; the stem is sometimes 34 feet in diameter, though the tree is rarely more than 50 or 60 feet high; it covers the sandy plains so entirely with its umbrella-shaped top, that a forest of these trees presents a compact surface, which at some distance seems to be a green field. Cape Verde has its name from the numbers that conceal the barren soil under their spreading tops; some of them are very old, and, with the dragon-tree at Teneriffe, are supposed to be the most ancient vegetable inhabitants of the earth. The pandanus candelabrum, instead of growing crowded together in masses like the boabab, stands solitary on the equatorial plains, with its lofty-forked branches ending in tufts of long stiff leaves. Numerous sedges, of which the papyrus is the most remarkable, give a character to this region, and cover boundless plains waving in the wind like corn-fields, while other places are overgrown by forests of gigantic grasses with branching stems.

A rich vegetation, consisting of impenetrable thickets of mangrove, the poisonous manchineel, and many large trees, cover the deltas of the rivers, and even grow so far into the water, that their trunks are coated with shell-fish, but the pestilential exhalations render it almost certain death to botanize in this luxuriance of nature.

Various trees of the soap and sapodilla tribes are peculiar to Africa; the butter tree of the enterprising but unfortunate Mungo Park, the star apple, the cream fruit, the custard apple, and the water vine, are plentiful in Senegal and Sierra Leone. The safu and bread-fruit of Polynesia are represented here by the musanga, a large tree of the nettle tribe, the fruit of which has the flavour of the hazel-nut. A few palms have very local habitations, as the elais Guineensis, found only on that coast. That graceful tribe is less varied in species in equatorial Africa than in the other continents.

The flora of south Africa differs entirely from that of the northern and tropical zones, and as widely from that of every other country, with the exception of Australia and some parts of Chili. The soil of the table-land at the Cape of Good Hope, stretching to an unknown distance, and of the Karoo plains and valleys between the mountains, is sometimes gravelly, but more frequently is composed of sand and clay; in summer it is dry and parched, and most of its rivers are dried up; it bears but a few stunted shrubs, some succulent plants and mimosas along the margin of the river courses. The sudden effect of rain on the parched ground is like magic; it is recalled to life, and in a short time is decked with a beautiful and peculiar vegetation, comprehending, more than any other country, numerous and distinctly defined foci of genera and species.

Twelve thousand species of plants have been collected in the colony of the Cape, in an extent of country about equal to Germany. Of these heaths and proteas are two very conspicuous tribes; there are 300 species of the former, and 200 of the latter, both of which have nearly the same limited range, though Mr. Bunbury found two heaths, and the protea cynaroides, the most splendid of the family, bearing a flower the size of a man's hat, on the hills round Graham's Town, in the eastern part of the colony. These two tribes of plants are so limited that there is not one of either to be seen north of the mountains which bound the Great Karoo, and by much the greatest number of them grew within 100 miles of Cape Town; indeed at the distance of only 40 miles the prevailing proteaceæ are different from those at the Cape. The leucadendron argenteum, or silver tree, which forms groves at the back of the table-mountain, is confined to the peninsula of the Cape. The beautiful disa grandiflora is found only in one particular place on the top of the Table-mountain.

The dry sand of the west coast and the country northward through many degrees of latitude is the native habitation of stapelias, succulent plants with square leafless stems and flowers like star-fish, with the smell of carrion. A great portion of the eastern frontier of the Cape colony and the adjacent districts are covered with extensive thickets of a strong succulent and thorny vegetation, called by the natives the bush: similar thickets occur again far to the west, on the banks of the river Gauritz. The most common plants of the bush are aloes of many species, all exceedingly fleshy and some beautiful: the great red-flowering arborescent aloe, and some others, make a conspicuous figure in the eastern part of the colony. Other characteristic plants of the eastern districts are the spek-boem, or portulacaria afra, schotia speciosa, and the great succulent euphorbias, which grow into real trees 40 feet high, branching like a candelabrum, entirely leafless, prickly, and with a very acrid juice. The euphorbia meloformis, whose bulb, three feet in diameter, lies on the ground, to which it is attached by slender fibrous roots, is confined to the mountains of Graaf Reynet. Euphorbias, in the Old World, correspond with the cactus tribe, which belong exclusively to the New. The zamia, a singular plant, having the appearance of a dwarf-palm without any real similarity of structure, belongs to the eastern districts, especially to the great tract of bush on the Caffir frontier.

Various species of acacia are indigenous and much circumscribed in their location: the acacia horrida, or the white-thorned acacia, is very common in the eastern districts and in Caffirland. The acacia cafra is strictly eastern, growing along the margins of rivers, to which it is a great ornament. The acacia detinens, or hook-thorn, is almost peculiar to Zand valley.

It appears from the instances mentioned, that the vegetation in the eastern districts of the colony differs from that on the western, yet many plants are generally diffused of orders and genera found only in this part of Africa. Nearly all the 300 species of the fleshy succulent tribe of mesembryanthemum, or Hottentots' fig; all the oxalis, or wood-sorrel tribe, except three in France and one in America; every species of gladiolus, with the exception of that in the corn-fields in Italy and France; ixias innumerable, one with petals of apple-green colour; geraniums, especially the genus pelargonium, or stork's bill, almost peculiar to this locality; many varieties of gnaphalium and xeranthemum; the brilliant strelitzia; 133 species of the house-leek tribe, all fleshy and leafless, attached to the soil by a single wiry root, and nourished from the atmosphere; diosmas are widely scattered in great variety; shrubby boragines with flowers of vivid colours, and orchideæ with large and showy blossoms. The leguminous plants and the cruciferæ of the Cape are peculiar; indeed all the vegetation has a distinct character, and both genera and species are confined within narrower limits than anywhere else, without any apparent cause to account for a dispersion so arbitrary.

Notwithstanding the peculiarity of character with which the botany of the Cape is so distinctly marked, it is connected with that of very remote countries by particular plants; for example, of the seven species of bramble which grow at the Cape, one is the common English bramble or blackberry. The affinity with New Holland is greater: in portions of the two countries in the same latitude there are several genera and species that are identical: proteas are common to both, so are several genera of irideæ, leguminosæ, ficoideæ, myrtaceæ, Banksias, diosmas, and some others. The botany of the Cape is connected with that of India, and even that of South America, by a few congeners.

The vegetation of Madagascar, though similar in many respects to the floras of India and Africa, nevertheless is its own: the brexiaceæ and chlenaceæ are orders found nowhere else; there are species of bignonia, cycades, and zamias, a few of the mangosteen tribe, and in the mountains some heaths. The hydrogeton fenestralis is a singular aquatic plant, with leaves like the dried skeletons of leaves, having no green fleshy substance, and the tanghinia veneniflua, which produces a poison so deadly that its seeds are used to execute criminals, and one seed is sufficient.

Some genera and species are common and peculiar to Madagascar, the Isle of Bourbon, and Mauritius; yet of the 161 known genera in Madagascar only 54 grow on the other two islands. The three islands are rich in ferns. The pandanus, or screw-pine genus, abounds in Bourbon and the Mauritius, where it covers sandy plains, sending off strong aërial roots from the stem, which strike into the ground and protect the plant from the violent winds. Of 290 genera in Bourbon and Mauritius, 196 also grow in India, though the species are different: there is also some resemblance to the vegetation of South Africa, and there is a solitary genus in common with America.

Eight or ten degrees north of Madagascar lies the group of the Seychelles Islands, in which are groves of the peculiar palm which bears the double cocoa-nut, or coco de mer, the growth of these islands only. Its gigantic leaves are employed in the construction of houses, and other parts of the plant are applied to various domestic purposes.

FLORA OF AUSTRALIA.

The interior of the Australian continent is so little known, that the flora which has come under observation is confined to a short distance from the coast; but it is of so strange and unexampled a character, that it might easily be mistaken for the production of another planet. Many entire orders of plants are known only in Australia, and the genera and species of others that grow elsewhere, assume new and singular forms. Evergreens, with hard narrow leaves of a sombre, melancholy hue, are prevalent, and there are whole shadowless forests of leafless trees, the foot-stalks dilated and set edgewise on the stem supply their place and perform the functions of nutrition: their inverted position gives them a singular appearance. Plants in other countries have glands on the under side of the leaves, but in Australia there are glands on both sides of these substitutes for leaves, which make them dull and lustreless, and the changes of the seasons have no influence on the unvarying olive-green of the Australian forests; even the grasses are separated from the gramineæ of other countries by a remarkable rigidity. Torres Straits, only 50 miles broad, separates this dry, sombre vegetation from the luxuriant jungle-clad shores of New Guinea, where deep and dark forests are rich in more than the usual tropical exuberance—a more complete and sudden change can hardly be imagined.

The peculiarly Australian vegetation is in the southern part of the continent of New Holland distributed in distinct foci in the same latitude, a circumstance of which the proteæ afford a remarkable instance. Nearly one half of the known species of these beautiful shrubs grow in the parallel of Port Jackson, from which they decrease in number both to the south and the north. In that latitude, however, there are twice as many species on the eastern side of the continent as there are on the western, and four times as many as in the centre. Although the proteas at both extremities of the continent have all the characters peculiar to Australia, yet those on the eastern coast resemble the South American species, while those on the western side have a resemblance to African forms, and are confined to the same latitudes.

Species of this genus are numerous in Van Diemen's Land, where they thrive at the elevation of 3500 feet, and also on the plains. The myrtle tribe form a conspicuous feature in Australian vegetation, particularly the genera eucalyptus, melaleuca, podocarpus and others, with splendid blossoms, white, purple, yellow, crimson: 100 species of the eucalypti, most of them large trees, grow in New Holland; they form great forests in the colony of Port Jackson. The leafless acacias, of which there are 93 species, are a prominent feature in the Australian landscape. The leaves, except in very young plants, are merely foliaceous foot-stalks, presenting their margin towards the stem, yet these and the eucalypti are the most leafy trees in the country. The genus casuarina, with its strange-jointed, drooping branches, called the march oak, holds a conspicuous place: they are chiefly confined to the principal parallel of this vegetation, and produce excellent timber; they grow also in the Malayan peninsula and South Sea Islands. The oxleya xanthoxyla, or yellow wood, one of the cedar tribe, grows to great size, and the podocarpus asplenifolia forms a new genus of the cone-bearing trees. Some of the nettle tribe grow 15 or even 20 feet high. The epacrideæ, with scarlet, rose, and white blossoms, supply the place of heaths, which do not exist here. The purple-flowering tremandreæ, the yellow-flowering dillenia, the doryanthis excelsa, the most splendid of the lily tribe, 24 feet high, with a brilliant crimson blossom, the Banksia, the most Australian of all the proteas, with zamias of new species, are all conspicuous in the vegetation of Port Jackson.

There is a change on the eastern coast of New Holland. The castanospermum Australe is so plentiful that it furnishes the principal food of the natives; a caper-tree of grotesque form, having the colossal dimensions of the Senegal boabab, and extraordinary trees of the fig genus, characterise this region. It sometimes occurs, when the seeds of these fig-trees are deposited by birds on the iron-bark tree, or eucalyptus resinifera, that they vegetate and enclose the trunk of the tree entirely with their roots, whence they send off enormous lateral branches, which so completely envelop the tree, that at last its top alone is visible in the centre of the fig-tree, at the height of 70 or 80 feet. The pandanus genus flourishes within the influence of the sea-air. There are only six species of palms, equally local in their habitations as elsewhere, not one of which grows on the west side of the continent. The araucaria excelsa, or Norfolk Island pine, produces the best timber of any tree in this part of Australia: it extends from the parallel of 29° on the east coast towards the equator, and grows over an area of 900 square miles, including New Norfolk, New Caledonia, and other islands, some of which have no other timber tree: it is supposed to exist only within the influence of the sea. The asphodelia abound and extend to the southern extremity of Van Diemen's Land.

The south-western districts of New Holland exhibit another focus of vegetation, less rich in species than that of Port Jackson, but not less peculiar. The Kingia Australis, or grass-tree, rises solitary on the sandy plains, with bare blackened trunks as if scathed by lightning, and tufts of long grassy leaves at their extremities; Banksias, particularly the kind called wild honeysuckle, are numerous; the stylidum, whose blossoms are even more irritable than the leaves of the sensitive mimosa, and plants with dry, everlasting blossoms, characterise the flora of these districts. The greater part of the southern vegetation vanishes on the northern coasts of the continent, and what remains is mingled with the cabbage-palm, various species of the nutmeg tribe, sandal-wood, and other Malayan forms, a circumstance that may hereafter be of importance to our colonists.

Orchideæ, chiefly terrestrial, are in great variety in the extratropical regions of New Holland, and the grasses amount to one-fourth of the monocotyledonous plants. Reeds of gigantic size form forests in the marshes, and kangaroo-grass covers the plains.

Beautiful and varied as the flora is, New Holland is by no means luxuriant in vegetation. There is little appearance of verdure, the foliage is poor, the forests often shadeless, and the grass thin; but in many valleys of the mountains, and even on some parts of the plains, the vegetation is vigorous. It is not the least remarkable circumstance in this extraordinary flora, that, with the exception of a few berries, there is no edible fruit, grain, or vegetable indigenous either in New Holland or Van Diemen's Land.

The plants of New Holland prevail in every part of Van Diemen's Land; yet the coldness of the climate and the height of the mountains permit genera of the northern hemisphere to be mixed with the vegetation of the country. Butter-cups, anemonies, and polygonums of peculiar species grow on the mountain tops, together with proteas and other Australian plants. The plains glow with the warm golden flowers of the black wattle, a mimosa emblematic of the island, and with the equally bright and orange blossom of the gorse, which perfumes the whole atmosphere. Only one tree-fern grows in this country: it rises 20 feet to the base of the fronds, which spread into an elegant top, producing a shadow gloomy as night-fall, and there are 150 species of orchis. The southern extremities both of New Holland and Van Diemen's Land are characterised by the prevalence of evergreen plants: but the trees here, as well as in the other parts of the southern hemisphere, do not shed their leaves periodically as with us.

The botany of New Zealand appears to be intimately allied to that of New Holland, South America, and South Africa, but chiefly to that of New Holland. Noble trees form impenetrable forests, sixty of which yield the finest timber, and many are of kinds to which we have nothing similar. Here there are no representatives of our oak, birch, or willow, but five species of beech and ten of pine have been discovered that are peculiar to the country. They are all alpine, and only descend to the level of the sea in the northern parts of the islands. The pines of the southern hemisphere are more local than in the northern; of the ten species peculiar to New Zealand it is not certain that more than two or three are found in the middle island, or that any of them grow south of the 40th parallel. The Kauri pine, or dammara australis, is indigenous in all the three islands, but it is the only cone-bearing tree in North Island, where it grows in hilly situations near the sea, shooting up with a clean stem 60 or 90 feet, sometimes 30 feet in diameter, with a spreading but thin top, and generally has a quantity of transparent yellow resin imbedded at its base. This fine tree does not grow beyond the 38° S. lat. The metrosideros tomentosa, with rich crimson blossoms, is one of the greatest ornaments of the forests, and the metrosideros robusta the most singular. It grows to a very great size, and sends shoots from its trunk and branches to the ground, which become so massive that they support the old stem which to all appearance loses its vitality; it is in fact an enormous epiphyte, growing to and not from the ground. Many of the smaller trees are of the laurel tribe, with poisonous berries. Besides there is the cabbage palm, the areca sapida, elder, the fuchsia excorticata, and other shrubs. Before New Zealand was colonized, the natives lived chiefly on the roots of the edible fern, pteris esculenta, with which the country is densely covered, mixed with a shrub that grows like a cypress, and the tea-plant, which is a kind of myrtle whose berries afford an intoxicating liquor. More than 90 species of fern are natives of these islands, some of which are arborescent and 40 feet high; the country is chiefly covered with these and with the New Zealand flax, phormium tenax, which grows abundantly both on the mountains and plains.

In Norfolk Island 152 species of plants are already known, and many, no doubt, are yet to be discovered. The araucaria excelsa and some palms are indigenous, and there are three times as many ferns as of all the other plants together.

The multitude of islands of Polynesia constitute a botanical region apart from all others, though it is but little varied, and characterised principally by the number of syngenesious plants with arborescent forms and tree-ferns. There are 50 varieties of the bread-fruit, which produce three or four crops in the year, and supply the natives with food, clothing, and timber; the cocoa-nut palm and the banana are on all the islands, and the pandanus, which thrives only when exposed to the sea-air. The tacca pinnatifida yields arrow-root; an intoxicating liquor is made from the fruit of one of the dracæna tribe, and the inner bark of the morus papyrifera is manufactured into cloth. Besides the cocoa-nut palm and pandanus, various trees grow on the coral islands, among others the fragrant suriana and the sweet-scented Tournefortia.

CHAPTER XXV.

AMERICAN VEGETATION—FLORA OF NORTH, CENTRAL, AND SOUTH AMERICA—ANTARCTIC FLORA—MARINE VEGETATION.

FROM similarity of physical circumstances the arctic flora of America bears a strong resemblance to that of the northern regions of Europe and Asia. This botanical district comprises Greenland, and extends considerably to the south of the arctic circle, especially at the eastern and western ends of the continent, where it reaches the 60th parallel of N. lat., and even more; it is continued along the tops of the Rocky Mountains almost to Mexico, and it re-appears on the White Mountains and a few other parts of the Alleganies.

Greenland has a much more arctic flora than Iceland; the valleys are entirely covered with mosses and marsh plants, and the gloomy rocks are cased in sombre lichens that grow under the snow, and the grasses on the pasture grounds that line the fiords are nearly four times less varied than those of Iceland. In some sheltered spots the service-tree bears fruit, and birches grow to the height of a few feet; but ligneous plants in general trail on the ground.

The arctic flora of America has much the same character with those of Europe and Asia, and many species are common to all; still more are representative, but there is a difference in the vegetation at the two extremities of the continent; there are 30 species in the east and 20 in the west end which grow nowhere else. The sameness of character changes with the barren treeless lands at the verge of the Arctic region, and the distribution of plants varies both with the latitude and the longitude. Taking a broad view of the botanical districts of North America, there are two woody regions, one on the eastern, the other on the western side of the continent, separated by a region of prairies where grasses and herbaceous plants predominate. The vegetation of these three parts, so dissimilar, varies with the latitude, but not after the same law as in Europe, for the winter is much colder, and the summer warmer, on the eastern coasts of America than on the western coast of Europe, owing in a great measure to the prevalence of westerly winds which bring cold and damp to our shores.

Boundless forests of black and white spruce with an undergrowth of reindeer moss cover the country south of the Arctic region, which are afterwards mixed with other trees; gooseberries, strawberries, currants, and some other plants thrive there. There are vast forests in Canada of pines, oak, ash, hiccory, red beech, birch, the lofty Canadian poplar, sometimes 100 feet high and 36 feet in circumference, and sugar maple; the prevailing plants are kalmias, azaleas, and asters, the former vernal, the latter autumnal; solidagos and asters are the most characteristic plants of this region.

The splendour of the North American flora is displayed in the United States; the American sycamore, chesnut, black walnut, hiccory, white cedar, wild cherry, red birch, locust-tree, tulip-tree or liriodendron, the glory of American forests, liquidambar, oak, ash, pine-trees of many species, grow luxuriantly with an undergrowth of rhododendrons, azaleas, andromedas, gerardias, calycanthus, hydrangea, and many more of woody texture, with an infinite variety of herbaceous and climbing plants.

The vegetation is different on the two sides of the Allegany Mountains; the locust-tree, Canadian poplar, hibiscus, and hydrangea, are most common on the west side; the American chesnut and kalmias are so numerous on the Atlantic side, as to give a distinctive character to the flora; here too aquatic plants are more frequent, among these the saracenia or pitcher-plant, singular in form, with leaves like pitchers covered with a lid, half-full of water.

The autumnal tints of the forests in the middle States are beautiful and of endless variety; the dark leaves of the evergreen pine, the red foliage of the maple, the yellow beech, the scarlet oak, and purple nyssa, with all their intermediate tints, ever changing with the light and distance produce an effect at sunset that would astonish the native of a country with a more sober-coloured flora under a more cloudy sky.

In Virginia, Kentucky, and the southern States the vegetation assumes a different aspect, though many plants of more northern districts are mixed with it. Trees and shrubs here are remarkable for broad shining leaves and splendid blossoms, as the gleditschia, catalpa, hibiscus, and all the family of magnolias, which are natives of the country, excepting a very few found in Asia and the Indian islands. They are the distinguishing feature of the flora from Virginia to the Gulf of Mexico, and from the Atlantic to the Rocky Mountains: the magnolia grandiflora and the tulip-tree are the most splendid specimens of this race of plants; the latter is often 120 feet high. The long-leaved pitch-pine, one of the most picturesque of trees, covers an arid soil on the coast of the Atlantic of 60,000 square miles. The swamps so common in the southern States are clothed with gigantic deciduous cypress, the aquatic oak, swampy hiccory, with the magnificent nelumbeum luteum and other aquatics, and among the innumerable herbaceous plants the singular dionæa muscipula, or American fly-trap; the trap is formed by two opposite leaves, covered with spines so irritable, that it instantly closes upon the insect that has come to suck its sweet juice. This magnolia region corresponds in latitude with the southern shores of the Mediterranean, but the climate is hotter and more humid, in consequence of which there is a considerable number of Mexican plants. A few dwarf-palms appear among the magnolias, and the forests in Florida and Alabama are covered with tillandsia usneoides, an air-plant, which hangs from the boughs.

Ten or twelve species of grass cover the extensive prairies or steppes of the valley of the Mississippi.

The forms of the Tartarian steppes appear to the north in the centaurea, artemisia, astragali, but the dahlias, œnotheras, with many more are their own. The helianthus and coreopsis, mixed with some European genera, mark the middle regions: and in the south, towards the Rocky Mountains, Clarkias and Bartonias are mixed with the Mexican genera of cactus and yucca. The western forest is less extensive and less varied than the eastern, but the trees are larger. This flora in high latitudes is but little known; the thuya gigantea on the Rocky Mountains and the coast of the Pacific is 200 feet high. Claytonias and currants, with plants of northern Asia, are found here.

Farther south the pinus Lambertiana is another specimen of the stupendous trees of this flora; seven species of it are indigenous in California, some of which have measured 200, and even 300 feet high, and 80 in circumference. Captain Belcher, in his 'Voyage on the Pacific,' mentions having measured an oak 27 feet in circumference, and another 18 feet girth at the height of 60 feet from the ground, before the branches began to spread. This is the native soil of the currant-bushes with red and yellow blossoms, of many varieties of lupins, peonies, poppies, and other herbaceous plants so ornamental in our gardens.

There are 332 genera of plants peculiar to North America, exclusive of Mexico, but no family of any great extent has yet been discovered there. About 160 large trees yield excellent timber; the wood of the pine-trees of the eastern forests is of inferior quality to that grown on the other side of the continent, and both appear to be less valuable than the pine-wood of Europe, which is best when produced in a cold climate. The pinus cembra and the pinus uncinata are the most esteemed of the Old World.

The native fruits of North America are mostly of the nut-kind, and there are many of these, to which may be added the Florida orange, the chicasa plum, the papaw, the banana, the red mulberry, and the plumlike fruit of the persimon. There are seven species of wild grapes, but good wine has not hitherto been produced. Although America has contributed so much to the ornament of our pleasure-grounds and gardens, yet there is not one North American plant which has become an object of extensive cultivation, while America has borrowed largely from other parts of the globe; the grapes cultivated in North America are European; tobacco, Indian corn, and many others of the utmost commercial value are strangers to the soil, having been introduced by the earliest inhabitants from Mexico and South America, which have contributed much more to general utility.

MEXICAN FLORA.

Mexico unites the vegetation of North and South America, though it resembles that of the latter more nearly. Whole provinces on the table-land and mountains produce alpine plants, oaks, chesnuts, and pines spontaneously. The edible-rooted nasturtium and the tuberous-rooted sorrel are peculiar.

The cheirostemon, or hand-tree, so named from the resemblance its stamens bear to the foot of a bird of prey, grows here, and also in the Guatimala forests.

The low lands of Mexico and Central America have a very rich flora, consisting of many orders and genera peculiar to them, and species without number, a great portion of which are unknown. The hymenea courbaril, from which the copal of Mexico is obtained, logwood, mahogany, and many other large trees, valuable for their timber, grow in the forests; sugar-cane, tobacco, indigo, aloe, yam, capsicum, and yucca are indigenous in Mexico and Central America. It is the native region of the melastomas, of which 620 species grow here; almost all the pepper tribe, the passifloræ, the ornament and pride of tropical America and the West Indian islands, begin to be numerous in these regions. The pine-apple is entirely American, growing in the woods and savannahs: it has been carried to the West Indies, to the East Indies and China, and is naturalized in all. This country has also produced the cherimoya, said to be the most exquisite of fruits. Hot arid tracts are covered with the cactus tribe, a plant of Central America and Mexico, which is more widely dispersed than the anana: some species bear a considerable degree of cold. They are social plants, inhabiting sandy plains in thickets, and of many species: their forms are various, and their blossoms beautiful. A few occur at a considerable distance from the tropics, to the north and the south. The night-flowering cereus grows in all its beauty in the arid parts of

Chili, filling the night air with its perfume. The cactus opuntia grows in the Rocky Mountains; and Sir George Back found a small island in the Lake of the Woods covered with it. This species has been brought to Europe, and now grows a common weed on the borders of the Mediterranean. In Mexico, the cochineal insect was collected from the cactus coccinellifer long before the Spanish conquest. There are large fields of American aloe, from which a liquor called pulque, and also an ardent spirit, are made. The ancient Mexicans made their hemp from this plant, and also their paper; and they used its thorns for nails. The forests of Panama contain at least 97 different kinds of trees, which grow luxuriantly in a climate where the torrents of rain are so favourable to vegetation, and so unfavourable to life that the tainted air is deadly even to animals. The flora of each West Indian island is similar to that of the continent opposite to it. The myrtus pimento, producing alspice, is common in the hills; cloves, nutmeg, custard-apple, guava, mango, the avocado pear, and tobacco are indigenous; the cabbage-palm grows to the height of 200 feet; the palma-real of Cuba is the most majestic of that noble family; and in Barbadoes there still exists a tree, but wearing out rapidly, which has given the island its name.

FLORA OF TROPICAL AMERICA.

Although the flora of tropical America is better known than that of Asia or Africa, there must still be thousands of plants of which we have no knowledge; and those which have come under observation are so varied and so numerous, that it is not possible to convey an idea of the peculiarities of this vegetation, or of the extent and richness of its woodlands. The upper Orinoco flows for some hundred miles chiefly through forests; and the selvas of the Amazons are six times the size of France. In these the trees are colossal, and the vegetation so matted together by underwood, creeping and parasitical plants, that the sun's rays can scarcely penetrate the dense foliage.

These extensive forests are by no means uniform; they differ on each side of the equator, though climate and other circumstances are the same. Venezuela, Guyana, the Amazons, and Brazil, are each the centre of a peculiar flora. So partial is this splendid vegetation, that almost each tributary of the great rivers has a flora of its own, particular families of plants are so restricted in their localities, and predominate so exclusively where they occur, that they change the appearance of the forest. Thus from the prevalence of the orders laurinæ, sapotaceæ, and others, which have leathery, shining, and entire leaves, the forests through which the Rio Negro, Cassiquiare, and Tuamine flow, differ in aspect from those of the other affluents of the Amazons. Even the grassy llanos, so uniform in appearance, have their centres of vegetation; and only agree with the pampas of Buenos Ayres in being covered with grass and herbs. In these tropical regions the flora varies with the altitude also. On the Andes, almost at the limit of vegetation, the ground is covered with purple, azure, and scarlet gentians, drabas, alchemillas, and many other brilliantly coloured alpine plants. This zone is followed by thickets of coriaceous-leaved plants, in perpetual bloom and verdure; and then come the forest-trees. Arborescent ferns ascend to 7000 feet; the coffee-tree and palms to 5000; and neither indigo nor cocoa can be cultivated lower than 2000.

Many parts of the coasts of Venezuela and Guiana are rendered pestilential by the effluvia of the mangrove, avicenna, and the manchineel, one of the euphorbia family, consisting of 562 species in tropical America, all having milky juice, deleterious in the greater number. The well-known poison curara is prepared by the Indians of Guiana from the fruit and bark of the bertholletia, of the order strychneæ, than which nature has probably produced no plants more deadly. The ourari is a creeping plant which yields the deadly wourali, the powerful effect of which was proved by Mr. Waterton's experiments.

The cinchona, or true bark-tree, grows only on the Cordilleras of the Andes. Medicinal qualities exist in other plants of different genera in Guiana, as the cusparia carony, which produces Angustura bark, and others with similar properties. The sapindus saponaria, or soap-tree, is used by the natives for washing. Capsicum, vanilla, the hoya, or incense plant, the dipterix odorat, whose fruit is the tonqua-bean, and the casada or mandioc, are natives of the country. The cow-tree, almost confined to the Cordillera of the coast of Venezuela, yields such abundance of nutritious milky juice that it is carried in pails like milk from the cow. The chocolate-palm, the cacao-shrub, fruits of the most excellent flavour, plants yielding balsam, resin, and gum, are numerous in the tropical regions. There the laurel tribe assume the character of majestic trees: some are so rich in oil, that it gushes from a wound in the bark. One of these laurels produces the essential oil which dissolves caoutchouc, or Indian rubber, used in rendering cloth waterproof.

Plantains of gigantic size form large forests; but palms are the most numerous and the most beautiful of all the trees in these countries. There are 90 species of them; and they are so local that a change takes place every 50 miles. They are the greatest ornament of the upper Orinoco.

The llanos of Venezuela and Guiana are covered with tall grass, mixed with lilies and other bulbous flowers, the sensitive mimosa, and palms constantly varying in species.

No language can describe the glory of the forests of the Amazons and Brazil, the endless variety of form, the contrasts of colour and size: there even the largest trees bear brilliant blossoms; scarlet, purple, blue, rose colour, and golden yellow, are blended with every possible shade of green. Majestic trees, as the bombax ceiba, the dark-leaved mora with its white blossoms, the fig, cashew and mimosa tribes, which are here of unwonted dimensions, and a thousand other giants of the forest, are contrasted with the graceful palm, the delicate acacia, reeds of a hundred feet high, grasses of 40, and tree-ferns in myriads. Passifloræ and slender creepers twine round the lower plants, while others as thick as cables climb the lofty trees, drop again to the ground, rise anew and stretch from bough to bough, wreathed with their own leaves and flowers, and studded with the vividly coloured blossoms of the orchideæ. An impenetrable and everlasting vegetation covers the ground; decay and death are concealed by the exuberance of life; the trees are loaded with parasites while alive; they become masses of living plants when they die.

One twenty-ninth part of the flowering plants of the Brazilian forests are of the coffea tribe, and the rose-coloured and yellow-flowering bignonias are among their greatest ornaments, where all is grace and beauty. Thousands of herbs and trees must still be undescribed, where each stream has its own vegetation. In those parts of Brazil less favoured by nature the forests consist of stunted deciduous trees, and the boundless plains have grasses, interspersed with myrtles and other shrubs.

The forests of Paraguay and Vermejo, in La Plata, are almost as rich as those of the tropics. Noble trees furnish timber and fruit; the algaroba, a kind of acacia, produces clusters of a bean, of which the Indians make bread, and also a strong fermented liquor; the palm and cinchona grow there; and the Yerba-maté, the leaves of which are universally used as tea in South America, and were in use before the Spanish conquest.

The sandy deserts towards the mountains are the land of the aloe and cactus in all their varieties. The fibres of the aloe are made into cordage by the Indians, for fishing-nets and other uses, and the juice affords them drink. Some larger species of cactus give durable wood ; and the cochineal insect, which feeds on them, is a valuable article of commerce.

Grass, clover, and the domiciled European and African thistles, with a solitary ombu at wide intervals, are the unvarying features of the pampas ; and thorny stunted bushes, characteristic of all deserts, are the only vegetation of the Patagonian shingle. But on the mountain valleys in the far south may be seen the winter's-bark, arbutus, new species of beech-trees, stunted berberries and misodendrons, which are singular kinds of parasitical plants.

Large forests of araucaria imbricata grow in the Andes of Chili and Patagonia. This tall and handsome pine, with cones the size of a man's head, supplies the natives with a great part of their food. It is said that the fruit of one large tree will maintain eighteen persons for a year.

Nothing grows under these great forests ; and when accidentally burnt down in the mountainous parts of Patagonia, they never rise again, but the ground they grew on is soon covered with an impenetrable bushwood of dwarf oak. In Chili the violently stinging loasa appears first in these burnt places, bushes grow afterwards, and then comes a tree grass, 18 feet high, of which the Indians make their huts. The new vegetation that follows the burning of primeval forests is quite unaccountable. The ancient and undisturbed forests of Pennsylvania have no undergrowth, and when burnt down they are succeeded by a thick growth of rhododendrons.

The southern coasts of Chili are very barren, and all plants existing there, even the herbaceous, have a tendency to assume a hard knotty texture. The stem of the wild potato, which is indigenous, becomes woody and bristly as it grows old. It is a native of the sea strand, and is never found more than 400 feet above it. In its wild state the root is small and bitter; it is one of many instances of the influence of cultivation in rendering unpromising plants useful to man.

Although the coast is barren, and the flora, at an elevation of 9000 feet on the Chilian Andes, almost identical with that of the Straits of Magellan, yet the climate is so mild in some valleys, especially that of Antuco, that the vegetation is semi-tropical. In it broad-leaved and bright-coloured plants, and the most fragrant and brilliant orchideæ, are mixed with the usual alpine genera. Dr. Poeppig says, that whatever South Africa or New Holland can boast of in beauty, in variety of form, or brilliancy of colour, is rivalled by the flora in the highest zone in this part of the Andes, even up to the region of perpetual snow; and, indeed, it bears a strong analogy to the vegetation of both these countries.

The humidity or dryness of the prevailing winds makes an immense difference in the character of the countries on each side of the Andes. In Peru they are bare of plants on the western side, while on the east there is exuberant vegetation; but it gradually disappears with the increasing height, till at an elevation of 13,782 feet arborescent plants vanish, and alpine races, of the most vivid beauty, succeed; which, in their turn, give place to the grasses at the height of 16,138 feet. Above these, in the dreary plains of Bombon, and other lands of the same altitude, even the thinly-scattered mosses are sickly; and at the height of 21,878 feet the snow-lichen forms the last show of vegetable life; confirming the observation of Don Ulloa, that the produce of the soil is the thermometer of Peru.

ANTARCTIC FLORA.

Terra del Fuego and Kerguelen's Islands are the northern boundary of the antarctic lands, which are scattered round the south pole at immense distances from one another. On these the vegetation decreases as the latitude increases, till at length utter desolation prevails; not a lichen covers the dreary storm-beaten rocks; not a sea-weed lives in the gelid waves. In the arctic regions, on the contrary, no land has yet been discovered that is entirely destitute of vegetable life. This remarkable difference does not so much depend on a greater degree of cold in winter as on the want of warmth in summer.

In the high northern latitudes, the power of the summer sun is so great as to melt the pitch between the planks of the vessels; while in corresponding southern latitudes Fahrenheit's thermometer does not rise above 14° at noon, at a season corresponding to our August. The perpetual snow comes to a much lower latitude in the southern lands than it does in the north. Sandwich Land, in a latitude corresponding to that of the north of Scotland, is perpetually covered with many fathoms of snow. A single species of grass, the aira antarctica, is the only flowering plant in the South Shetland Islands, which are no less ice-bound; and Cockburn Island, one of that group, in the 60th parallel, contains the last vestiges of vegetation; while the namesake islands, in an equally high latitude, to the north of Scotland, are inhabited and cultivated; nay, South Georgia, in a latitude similar to that of Yorkshire, is always clad in frozen snow, and only produces some mosses, lichens, and wild burnet; while Iceland, 10 degrees nearer the pole, has 870 species, more than half of which are flower-bearing.

The forest-covered islands of Tierra del Fuego are only 360 miles from the desolate Shetland group. Such is the difference that a few degrees of latitude can produce in these antarctic regions, combined with an equable climate and excessive humidity. The prevalence of evergreen plants is the most characteristic feature in the Fuegian flora. Densely entangled forests of winter's bark, and two species of beech trees, grow from the shore to a considerable height on the mountains. Of these, the fagus deltoides, which never loses its brownish green leaves, prevails almost to the exclusion of the evergreen winter's bark and the deciduous beech, which is very beautiful. There are dwarf species of arbutus, the myrtus nummularia, which is used instead of tea, besides berberry, currant, and fuchsia. Peculiar species of ranunculi, calceolarias, caryophylleæ, cruciform plants and violets. Wild celery and scurvy grass are the only edible plants; and a bright yellow fungus, which grows on the beech trees, forms a great part of the food of the natives. There is a greater number of plants in Tierra del Fuego, either identical with those in Great Britain, or representatives of them, than exists in any other country in the southern hemisphere. The sea-pink, or thrift, the common sloewort, primula hirsuta, and at least thirty other flowering plants, with almost all the lichens, 48 mosses, and many other plants of the cyptogamous kinds, are identically the same; while the number of genera common to both countries is still greater, and though unknown in the intermediate latitudes, reappear here. Hermite Island, west from Cape Horn, is a forest land, covered with winter's bark and the Fuegian beeches; and is the most southern spot on earth on which arborescent vegetation is found. An alpine flora, many of them of European genera, grows on the mountains; succeeded higher up by mosses and lichens. Mosses are exceedingly plentiful throughout Fuegia; but they abound in Hermite Island more than in any other country, of singular and beautiful kinds.

Although the Falkland Islands are in a lower latitude than Tierra del Fuego, not a tree is to be seen. The veronica elliptica, resembling a myrtle, which is extremely rare, and confined to West Falkland, is the only large shrub; a white flowering plant, like an aster, about four feet high, is common; while a bramble, a crowberry, and a myrtle, bearing no resemblance, however, to the European species, trail on the ground, and afford edible fruit. The balsam bog, or bolax glebaria, and grasses, form the only conspicuous feature in the botany of these islands; and, together with rushes and Dactylis Cæspitosa, or Tussack grass, cover them, almost to the exclusion of other plants. The bolax grows in tufted hemispherical masses, of a yellow-green colour, and very firm substance, often four feet high, and as many in diameter, from whence a strong-smelling resinous substance exudes perceptible at a distance. This plant has umbelliferous flowers, and belongs to the carrot order, but forms an antarctic genus quite peculiar.

The tussock grass is the most useful and the most singular plant in this flora. It covers all the small islands of the group, like a forest of miniature palm-trees, and thrives best on the shores exposed to the spray of the sea. Each tussock is an isolated plant, occupying about two square yards of ground. It forms a hillock of matted roots, rising straight and solitary out of the soil, often six feet high and four or five in diameter; from the top of which it throws out a thick grassy foliage of blades, six feet long, drooping on all sides, and forming with the leaves of the adjacent plants an arch over the ground beneath, which yields shelter to sea-lions, penguins, and petrels. Cattle are exceedingly fond of this grass, which yields annually a much greater supply of excellent fodder than the same extent of ground would do either of common grass or clover. Both the tussock grass and the bolax are found, though sparingly, in Tierra del Fuego; indeed, the vegetation of the Falkland Islands consists chiefly of the mountain plants of that country, and of those that grow on the arid plains of Patagonia; but it is kept close to the ground by the fierceness of the terrific gales that sweep over these antarctic islands. Peculiar species of European genera are found here, as a calceolaria, wood sorrel, and a yellow violet; while the shepherd's purse, cardamine hirsuta, and the primula farinosa, appear to be identical with those at home. In all there are scarcely 120 flowering plants, including grasses. Ferns and mosses are few, but lichens are in great variety and abundance, among which many are identical with those in Britain.

In the eastern hemisphere, far, far removed from the Falkland group, the Auckland Islands lie in the boisterous ocean south of New Zealand. They are covered with dense and all but impenetrable thickets of stunted trees, or rather shrubs, about 20 or 30 feet high, gnarled by gales from a stormy sea. There is nothing analogous to these shrubs in the northern hemisphere; but the veronica elliptica, a native of Tierra del Fuego and New Zealand, is one of them. Fifteen species of ferns find shelter under these trees, and their fallen trunks are covered with mosses and lichens. Eighty flowering plants were found during the stay of the discovery ships, of which fifty-six are new; and half of the whole number are peculiar to this group and to Campbell's Island. Some of the most beautiful flowers grow on the mountains, others are mixed with the ferns in the forests. A beautiful plant was discovered, like a purple aster, a veronica, with large spikes of ultramarine colour; a white one, with a perfume like jessamine; a sweet-smelling alpine hierochloe; and in some of the valleys the fragrant and bright-yellow blossoms of a species of asphodel were so abundant that the ground looked like a carpet of gold. A singular plant grows on the sea-shore, having bunches of green waxy blossoms the size of a child's head. There are also antarctic species of European genera, as beautiful red and white gentians, geraniums, &c. The vegetation is characterised by an exuberance of the finer flowering plants, and an absence of grasses and sedges; but the landscape, though picturesque, has a sombre aspect, from the prevalence of brownish-leaved plants of the myrtle tribe.

Campbell's Island lies 120 miles to the south of the Auckland group, and is much smaller, but from the more varied form of its surface it is supposed to produce as many species of plants. During the two days the discovery ships, under the command of

Sir James Ross, remained there, between 200 and 300 were collected, of which 66 were flowering plants, 14 of which were peculiar to the country. Many of the Auckland Island plants were found here, yet a great change had taken place; 34 species had disappeared and were replaced by 20 new, all peculiar to Campbell's Island alone, and some were found that hitherto had been supposed to belong to Antarctic America only. In the Auckland group only one-seventh of the plants are common to other Antarctic lands, whilst in Campbell's Island a fourth are natives of other longitudes in the Antarctic Ocean. The flora of Campbell's Island and the Auckland group is so intimately allied to that of New Zealand, that it may be regarded as the continuation of the latter, under an Antarctic character, though destitute of the beech and pine trees. There is a considerable number of Fuegian plants in the islands under consideration, though 4000 miles distant, and whenever their flora differs in the smaller plants from that of New Zealand, it approximates to that of Antarctic America; but the trees and shrubs are entirely dissimilar. The relation between this vegetation and that of the northern regions is but slight. The Auckland group and Campbell's Island are in a latitude corresponding to that of England, yet only three indigenous plants of our island have been found in them, namely, the cardamine hirsuta, montia, and callitriche. This is the utmost southern limit of tree-ferns.

Perhaps no spot in either hemisphere, at the same

distance from the pole, is more barren than Kerguelen Islands, lying in a remote part of the south polar ocean. Only 18 species of flowering plants were found there, which is less than the number in Melville Island, in the Arctic Seas; and three times less than the number even in Spitzbergen. The whole known vegetation of these islands only amounts to 150, including sea-weeds. The pringlea, a kind of cabbage, acceptible to those who have been long at sea, is peculiar to the island, and grass, together with a plant similar to the bolax of the Falkland Islands, covers large tracts. About 20 mosses, lichens, &c. are only found in these islands, but many of the others are also native in the European Alps and north polar regions. It is a very remarkable circumstance in the distribution of plants, that there should be so much analogy between the floras of places so far apart as Kerguelen Islands, the groups south from New Zealand, the Falkland Islands, South Georgia, and Tierra del Fuego.

MARINE VEGETATION.

A vegetable world lies hid beneath the surface of the ocean, altogether unlike that on land, and existing under circumstances totally different with regard to light, heat, and pressure, yet sustained by the same means. Carbonic acid and ammonia are as essential, and metallic oxides are as indispensable to marine vegetation as they are to land plants. Sea water contains ammonia, and something more than a twelve-thousandth part of its weight of carbonate of lime, yet that minute portion is sufficient to supply all the shell-fish and coral insects in the sea with materials for their habitations, as well as food for vegetation. Marine plants are more expert chemists than we are, for the water of the ocean contains rather less than a millionth part of its weight of iodine, which they collect in quantities, impossible for us to obtain otherwise than from their ashes.

Sea-weeds fix their roots to any thing; to stone, wood, and to other sea-weeds: they must therefore derive all their nourishment from the water, and the air it contains; and the vital force or chemical energy by which they decompose and assimilate the substances fit for their maintenance, is the sun's light.

Flower-bearing sea-weeds are very limited in their range, which depends upon the depth of water and the nature of the coasts, but the cryptogamic kinds are widely dispersed, some species are even found in every climate from pole to pole.

Marine vegetation varies both horizontally and vertically with the depth, and it seems to be a general law throughout the ocean that the light of the sun and vegetation end together; it consequently depends on the power of the sun and the transparency of the water; so different kinds of sea-weeds affect different depths, where the weight of the water, the quantity of light and heat, suit them best. One great marine zone lies between the high and low water marks, and varies in species with the nature of the coasts, but exhibits similar phenomena throughout the northern hemisphere. In the British seas, where, with two exceptions, the whole flora is cryptogamic, this zone does not extend deeper than 30 fathoms, but is divided into two distinct provinces, one to the south and another to the north. The former includes the southern and eastern coasts of England, the southern and western coasts of Ireland, and both the channels; while the northern flora is confined to the Scottish seas and the adjacent coasts of England and Ireland. The second British zone begins at low-water mark, and extends below it to a depth from 7 to 15 fathoms. It contains the great tangle sea-weeds, growing in miniature forests, mixed with fuci, and is the abode of a host of animals. The nulliflora, a coral-like sea-weed, is the last plant of this zone, and the lowest in these seas, where it does not extend below the depth of 60 fathoms, but in the Mediterranean it is found at 70 or 80 fathoms, and is the lowest plant in that sea. The same law prevails in the Bay of Biscay, where one set of sea-weeds is never found lower than 20 feet below the surface; another only in the zone between the depths of 5 and 30 feet; and another between 15 and 35 feet. In these two last zones they are most numerous; at a greater depth the kinds continue to vary, but their numbers decrease. The seeds of each kind float at the depth most genial to the future plant, they must therefore be of different weights. The distribution in the Egean Sea was found by Professor E. Forbes to be perfectly similar, only that the vegetation is different, and extends to a greater depth in the Mediterranean than in more northern seas. He also observed that sea-weeds growing near the surface are more limited in their distribution than those that grow lower down, and that with regard to vegetation depth corresponds with latitude, as height does on land. Thus the flora at great depths, in warm seas, is represented by kindred forms in higher latitudes. There is every reason to believe that the same laws of distribution prevail throughout the ocean and every sea.

Two genera of weeds inhabit the sea: a jointed kind, which includes the confervæ, which are plants having a thread-like form; and the jointless kind, to which belong dulse, laver, the kinds used for making kelp, vegetable glue, and iodine; that in the Indian Archipelago of which the sea-swallow makes the edible nests; and all the huge species which grow in submarine forests, or float like green meadows in the open ocean.

Sea-weeds adhere firmly to the rocks before their fructification, but they are easily detached afterwards, which accounts for some of the vast fields of floating weeds; but others, of gigantic size and wide distribution, are supposed to grow unattached in the water itself. There are permanent bands of sea-weed in our British Channel and in the North Sea, of the kind called felum, which grows abundantly on the western coasts of the Channel, and they lie in the direction of the currents, in beds 15 or 20 miles long, and not more than 600 feet wide. These bands must oscillate with the tides between two corresponding zones of rest, one at the turn of the flood, and the other at the turn of the ebb. It is doubtful whether the fucus natans or sargassum bucciforum grows on rocks at the bottom of the Atlantic, between the parallels of 40° north and south of the equator, and when detached, is drifted uniformly to particular spots which never vary; or whether it is propagated and grows in the water; but the mass of that plant, west of the Azores, occupies an area equal to that of France, and has not changed its place since the time of Columbus. Fields of the same kind cover the sea at the Bahama Islands and other places, and two new species of it were discovered in the Antarctic seas.

The macrocystis pyrifera and the laniaria radiata are the most remarkable of marine plants for their gigantic size and the extent of their range. They were met with on the Antarctic coasts, two degrees nearer the south pole than any other vegetable production, forming the utmost limit of vegetable life in the south polar seas. The macrocystis pyrifera exists in vast detached masses, like green meadows, in every latitude from the south polar ocean to the 45th degree N. lat. in the Atlantic, and to the shores of California in the Pacific, where there are fields of it so impenetrable, that it has saved vessels driven by the heavy swell towards that shore from shipwreck. It is never seen where the temperature of the water is at the freezing-point, and is the largest of the vegetable tribe, being occasionally

300 or 400 feet long. The laminaria abounds off the Cape of Good Hope and in the Antarctic Ocean. These two species form great part of a band of sea-weed that girds Kerguelen Islands so densely, that a boat can scarcely be pulled through it, and they are found in great abundance on the coasts of the Falkland group, and also in vast fields in the open sea, hundreds of miles from any land; had it ever grown on the distant shores, it must have taken ages to travel so far, drifted by the wind, currents, and the sand of the seas. The red, green, and purple lavers of Great Britain are found on the coasts of the Falkland Islands, and though some of the northern weeds are not found in the intervening warm seas, they re-appear here. The lessonia is the most remarkable marine plant in this group of islands. Its stems, much thicker than a man's leg, and from 8 to 10 feet long, fix themselves by clasping fibres to the rocks beyond the high-water mark. Many branches shoot upwards from these stems, from which long leaves droop into the water like willows. There are immense submarine forests off Patagonia and Tierra del Fuego, attached to the rocks at the bottom. These plants are so strong and buoyant, that they bring up large masses of stone, and as they grow slanting, and stretch along the surface of the sea, they are sometimes 300 feet long. The quantity of living creatures which inhabit these marine forests and the parasitical weeds attached to them is inconceivable, they absolutely teem with life.

Great patches of confervæ are occasionally met with in the high seas. Bands several miles long, of a reddish brown species, like chopped hay, occur off Bahia, on the coast of Brazil; the same plant is said to have given the name to the Red Sea; and different species are common in the Australian seas.

CHAPTER XXVI.

DISTRIBUTION OF INSECTS.

THREE hundred thousand insects are known: some with wings, others without; some are aquatic, others are aquatic only in the first stage of their existence; and many are parasitical. Some land insects are carnivorous, others feed on vegetables; some of the carnivorous tribe live on dead, others on living animals, but they are not half so numerous as those that live on vegetables. Some change as they are developed; in their first stage they eat animal food, and vegetables when they come to maturity. Insects increase in kinds and in numbers from the poles to the equator: in a residence of eleven months in Melville Island, Sir Edward Parry found only six species, because lichens and mosses do not afford nourishment for the insect tribes, though it is probable that every other kind of plant gives food and shelter to more than one species; it is even said that 40 different insects are quartered upon the common nettle.

The increase of insects from the poles to the equator does not take place at the same rate everywhere. The polar regions and New Holland have very few specifically and individually; they are more abundant in North Africa, Chili, and the sandy deserts west from Brazil; North America has fewer species than Europe in the same latitude, and Asia has few varieties of species in proportion to its size; Europe, especially Germany, produces many more species than intertropical Africa; Caffraria, the African and Indian islands, are nearly the same as to species; but by far the richest of all, both in species and numbers, are central and intertropical America. Beetles are an exception to the law of increase towards the equator, as they are infinitely more numerous in species in the temperate regions of the northern hemisphere than in equatorial countries. The location of insects depends upon that of the plants which yield their food; and as almost each plant is peopled with inhabitants peculiar to itself, insects are distributed over the earth in the same manner as vegetables; the groups consequently are often confined within narrow limits, and it is extraordinary that, notwithstanding their powers of locomotion, they often remain within a particular compass, though the plants, and all other circumstances in their immediate vicinity, appear equally favourable for their habitation.

Though insects are distributed in certain limited groups, yet most of the families have representatives in all the great regions of the globe, and some identical species are inhabitants of countries far from one another. The venussi cardui live in all the four quarters of the globe and in Australia; and one, which never could have been conveyed by man, is native in southern Europe, the coast of Barbary, and Chili. It is evident from these circumstances that not only each group, but also each particular species, must have been originally created in the places they now inhabit.

Mountain-chains are a complete barrier to insects, even more so than rivers: not only lofty mountains like the Andes divide the kinds, but they are even different on the two sides of the Col de Tende in the Alps. Each soil has kinds peculiar to itself, whether dry or moist, cultivated or wild, meadow or forest. Stagnant water and marshes are generally full of them; some live in water, some run on its surface, and every water-plant affords food and shelter to many different kinds. The east wind seems to have considerable effect in bringing the insect or in developing the eggs of certain species; for example, the aphis, known as the blight in our country, lodges in myriads on plants, and shrivels up their leaves after a continued east wind.

Temperature, by its influence on vegetation, has an indirect effect on the insects that are to feed upon plants, and extremes of heat and cold have more influence on their locality than the mean annual temperature. Thus in the polar regions the musquito tribes are more numerous and more annoying than in temperate countries, because they pass their early stages of existence in water, which shelters them, and the short but hot summer is genial to their brief span of life.

In some instances height corresponds with latitude. The parnassus apollo, a butterfly native in the plains of Sweden, is also found in the Alps, the Pyrenees, and even on the Himalaya. Some insects require several years to arrive at their perfect state. They lie buried in the ground in the form of grubs: the cockchafer comes to maturity in three years, and some American species require a much longer time.

Insects do not attain their perfect state till the plants they are to feed upon are ready for them. Hence in cold and temperate climates their appearance is simultaneous with vegetation; and as the rainy and dry seasons within the tropics correspond to our winter and summer, insects appear there after the rains and vanish in the heat: the rains, if too violent, destroy them; and in countries where that occurs, there are two periods in the year in which they are most abundant, one before and one after the rains. It is also observed in Europe that insects decrease in the heat of summer and become more numerous in autumn: the heat is thought to throw some into a state of torpor, but the greater number perish.

It is not known that any insect depends entirely upon only one species of plant for its existence, or whether it may not have recourse to congeners should its habitual plant perish. When particular species of plants of the same family occur in places widely apart, insects of the same genus will be found on them, so that the existence of the plant may often be inferred from that of the insect, and in several instances the converse.

When a plant is taken from one country to another in which it has no congeners, it is not attacked by the insects of the country: thus our cabbages and carrots in Cayenne are not injured by the insects of that country, and the tulip-tree and other magnolias are not molested by our insects; but if a plant has congeners in its new country, the inhabitants will soon find their way to the stranger.

The common fly is one of the most universal of insects, yet it was unknown in some of the South Sea Islands till it was carried there by ships from Europe, and it has now become a plague.

The musquito and culex are spread over the world more generally than any other tribe: it is the torment of men and animals from the poles to the equator by night and by day: the species are numerous and their location partial. In the Arctic regions the culex pipiens, which passes two-thirds of its existence in water, swarms in summer in myriads: the lake Myvatr, in Iceland, has its name from the legions of these tormentors that cover its surface. They are less numerous in middle Europe, though one species of musquito, the simulia columbaschensis, which is very small, appears in such clouds in parts of Hungary, especially the bannat of Temeswar, that it is not possible to breathe without swallowing many: even cattle and children have died from them. In Lapland there is a plague of the same kind. Of all places on earth the Orinoco and other great rivers of tropical America are the most obnoxious to this plague. The account given by

Baron Humboldt is really fearful: at no season of the year, at no hour of the day or night, can rest be found; whole districts in the Upper Orinoco are deserted on account of these insects. New species follow one another with such precision, that the time of day or night may be known accurately from their humming noise, and from the different sensations of pain which the different poisons produce. The only respite is the interval of a few minutes between the departure of one gang and the arrival of their successors, for the species do not mix. On some parts of the Orinoco the air is one dense cloud of poisonous insects to the height of 20 feet. It is singular that they do not infest rivers that have black water, and each white stream is peopled with its own kinds; though ravenous for blood, they can live without it, as they are found where no animals exist.

In Brazil the quantity of insects is so great in the woods, that their noise is heard in a ship at anchor some distance from the shore.

Various genera of butterflies and moths are very limited in their habitations, others are dispersed over the world, but the species are almost always different. Bees and wasps are equally universal, yet each country has its own. The common honey-bee is the only European insect directly useful to man; it was introduced into North America not many years ago, and is now spread over the continent. European bees, of which there are many species, generally have stings; the Australian bee, like a black fly, is without a sting; and in Brazil there are 30 species of stingless bees.

Fire-flies are mostly tropical, yet there are four species in Europe; in South America there are three species, and so brilliant that their pale green light is seen at the distance of 200 paces.

The silkworm comes from China, and the cochineal insect is a native of South and Central America: there are many species of it in other countries. The coccus lacca is Indian; the coccus ilicis lives in Italy, and there is one in Poland, but neither of these have been cultivated.

Scorpions under various forms are in all warm climates; 24 species are peculiar to Europe, but they are small in comparison with those in tropical countries: one in Brazil is six inches long. As in musquitoes, the poison of the same species is more active in some situations than in others. At Cumana the sting of the scorpion is little feared, while that of the same species in Carthagena causes loss of speech for many days.

Ants are universally distributed, but of different kinds: they are so destructive in South America, that Baron Humboldt says there is not a manuscript in that country a hundred years old. Near great rivers they build their nests above the line of the annual inundations.

Spiders abound more in Europe than elsewhere; of 900 known kinds, each country has its own, varying in size, colour, and habits, from the huge bird-catching spider of South America, to the almost invisible European gossamer floating in the air on its silvery thread. Many of this ferocious family are aquatic; and spiders, with some other insects, are said to be the first inhabitants of new islands.

The migration of insects is one of the most curious circumstances relating to them: they sometimes appear in great flights in places where they never were seen before, and they continue their course with perseverance which nothing can check. This has been observed in the migration of crawling insects: caterpillars have attempted to cross a stream. Countries near deserts are most exposed to the invasion of locusts, which deposit their eggs in the sand, and when the young are hatched by the sun's heat, they emerge from the ground without wings; but as soon as they attain maturity, they obey the impulse of the first wind and fly, under the guidance of a leader, in a mass, whose front keeps a straight line, so dense that it forms a cloud in the air, and the sound of their wings is like the murmur of the distant sea. They take immense flights, crossing the Mozambique channel from Africa to Madagascar, which is 120 miles broad: they come from Barbary to Italy, and a few have been seen in Scotland. Even the wandering tribes of locusts differ in species in different deserts, following the universal law of organised nature. Mr. Ehrenberg has discovered a new world of creatures in the Infusoria, so minute that they are invisible to the naked eye. He found them in fog, rain, and snow, in the ocean, in stagnant water, in animal and vegetable juices, in the dusty air that sometimes falls on the ocean; and he detected 18 species 20 feet below the surface of the ground, in peat earth, which was full of microscopic live animals: they exist in ice, and are not killed by boiling water. This lowest order of animal life is much more abundant than any other, and new species are found every day. Magnified, some of them seem to consist of a transparent vesicle, and some have a tail: they move with great alacrity, and show intelligence by avoiding obstacles in their course: others have siliceous shells. Language, and even imagination, fails in the attempt to describe the inconceivable myriads of these invisible inhabitants of the ocean, the air, and the earth: they no doubt become the prey of larger creatures, and perhaps blood-sucking insects may have recourse to them when other prey is wanting.

CHAPTER XXVII.

DISTRIBUTION OF FISHES, AND OF THE MARINE MAMMALIA, PHOCÆ, DOLPHINS, AND WHALES.

Before Sir James Ross's voyage to the Antarctic regions, the profound and dark abysses of the ocean were supposed to be entirely destitute of animal life; now it may be presumed that no part of it is uninhabited, since during that expedition live creatures were fished up from a depth of 6000 feet. But as most of the larger fish usually frequent shallow water near the coasts, deep seas must form barriers as impassable to the greater number of them as mountains do to land animals. The polar, the equatorial ocean, and the inland seas, have each their own particular inhabitants; almost all the species and many of the genera of the marine creation are different in the two hemispheres, and even in each particular sea; and under similar circumstances the species are for the most part representative, not the same. Identity of species, however, does occur, even at the two extremities of the globe, for living animals were brought up from the profound depths of the Antarctic Ocean which Sir James Ross recognised to be the very same species which he had often met with in the Arctic seas. "The only way they could have got from the one pole to the other must have been through the tropics; but the temperature of the sea in these regions is such that they could not exist in it unless at a depth of nearly 2000 fathoms. At that depth they might pass from the Arctic to the Antarctic Ocean without a variation of 5 degrees of temperature; whilst any land animal, at the most favourable season, must experience a difference of 50 degrees, and if in winter, no less than 150 degrees of Fahrenheit's thermometer;"—a strong presumption that marine creatures can exist at the depth and under the enormous pressure of 12,000 feet of water.

The form and nature of the coasts have great influence on the distribution of fishes; when they are uniformly of the same geological structure, so as to afford the same food and shelter, the fish are the same, or similar.

The ocean, the most varied and most wonderful part of the creation, absolutely teems with life: "things innumerable, both great and small, are there." The forms are not to be numbered even of those within our reach; yet, numerous as they are, few have been found exempt from the laws of geographical distribution.

The discoloured portions of the ocean generally owe the tints they assume to myriads of insects. In the Arctic seas, where the water is pure transparent ultramarine colour, parts of 20 or 30 square miles, 1500 feet deep, are green and turbid from the quantity of minute animalcules. Captain Scoresby calculated that it would require 80,000 persons, working unceasingly from the creation of man to the present day, to count the number of insects contained in two miles of the green water. What then must be the amount of animal life in the polar regions, where one-fourth part of the Greenland sea, for 10 degrees of latitude, consists of that water. These animalcules are of the medusa tribe, mixed with others that are moniliform. Some medusæ are very large, floating like jelly; and although apparently carried at random by the waves, each species has its definite location, and even locomotion. One species comes in spring from the Greenland seas to the coast of Holland; and Baron Humboldt met with an immense shoal of them in the Atlantic, migrating at a rapid rate.

Dr. Pœppig mentions a stratum of red water near Cape Pelaris, 24 miles long and 7 broad, which seen from the mast-head appeared dark-red, but on proceeding it became a brilliant purple, and the wake of the vessel was rose-colour. The water was perfectly transparent, but small red dots could be discerned moving in spiral lines. The vermilion sea off California is no doubt owing to a similar cause, as Mr. Darwin found red and chocolate-coloured water on the coast of Chili over spaces of several square miles full of microscopic animalcules, darting about in every direction, and sometimes exploding. Infusoria are not confined to water; the bottom of the sea swarms with them. Siliceous-coated infusoria are found in the mud of the coral islands under the equator; and 68 species were discovered in the mud in Erebus Bay, near the Antarctic pole. These minute forms of organised being, invisible to the naked eye, are intensely and extensively developed in both of the polar oceans, and serve for food to the higher orders of fish in latitudes beyond the limits of vegetation. Some are peculiar to each of the polar seas, some are common to both, and a few are distributed extensively throughout the ocean.

The enormous prodigality of animal life supplies the place of vegetation, so scanty in the ocean in comparison with that which clothes the land, and which probably would be insufficient for the supply of the marine creation, were the deficiency not made up by the superabundant land vegetation, and insects carried to the sea by rivers. The fish that live on sea-weed must bear a smaller proportion to those that are predacious, than the herbivorous land animals do to the carnivorous. Fish certainly are most voracious; none are without their enemies; they prey and are preyed upon; and there are two which devour even the live coral, hard as its coating is; nor does the coat of mail of shell-fish protect them. Whatever the proportion may be which predatory fish bear to herbivorous, the quantity of both must be enormous, for, besides the infusoria, the great forests of fuci and sea-weed are everywhere a mass of infinitely varied forms of being, either parasitical, feeding on them, seeking shelter among them, or in pursuit of others.

The observations of Professor E. Forbes in the Egean Sea show that depth has great influence in the geographical distribution of marine animals. From the surface to the depth of 230 fathoms there are eight distinct regions in that sea, each of which has its own vegetation and inhabitants. The number of shell-fish and other marine animals is greater specifically and individually between the surface and the depth of two fathoms than in all the regions below taken together, and both decrease downwards to the depth of 105 fathoms; between which and the depth of 230 only eight shell-fish were found; and animal life ceases in that part of the Mediterranean at 300 fathoms. The changes in the different zones are not abrupt; some of the creatures of an under region always appear before those of the region above vanish; and although there are a few species the same in some of the eight zones, only two are common to all. Those near the surface have forms and colours belonging to the inhabitants of southern latitudes, while those lower down are analogous to the animals of northern seas; so that in the sea depth corresponds with latitude, as height does on land. Moreover, the extent of the geographical distribution of any species is proportional to the depth at which it lives; consequently, those living near the surface are less widely dispersed than those inhabiting deep water. Professor Forbes also discovered several shell-fish living in the Mediterranean that have hitherto only been known as fossils of the tertiary strata; and also that the species least abundant as fossils are most numerous alive, and the converse; hence the former are near their maximum, while the latter are approaching to extinction. These very important experiments, it is true, were confined to the Mediterranean; but analogous results have been obtained in the Bay of Biscay and in the British seas. There are four zones of depth in our seas, each of which has its own inhabitants, consisting of shell-fish, crustaceæ, corallines, and other marine creatures. The first zone lies between high and low-water marks, consequently it is shallow in some places and 30 feet deep in others. In all parts of the northern hemisphere it presents the same phenomena; but the animals vary with the nature of the coast, according as it is of rock, gravel, sand, or mud. In the British seas the animals of this littoral or coast zone are distributed in three groups that differ decidedly from one another, though many are common to all. One occupies the seas on the southern shores of our islands and both channels; a middle group has its centre in the Irish seas; and the third is confined to the Scottish seas, and the adjacent coasts of England and Ireland. The second zone extends from the low-water mark to a depth below it of from 7 to 15 fathoms, and is crowded with animals living on and among the sea-weeds, as radiated animals, shell-fish, and many zoophytes. In the third zone, which is below that of vegetable life, marine animals are more numerous and of greater variety than in any other. It is particularly distinguished by arborescent creatures, that seem to take the place of plants, carnivorous mollusca, together with large and peculiar radiata. It ranges from the depth of 15 to 50 fathoms. The last zone is the region of stronger corals, peculiar mollusca, and of others that only inhabit deep water.

This zone extends to the depth of 100 fathoms or more.

Except in the Antarctic seas, the superior zone of shell-fish is the only one of which anything is known in the great oceans, which have numerous special provinces; but, according to Mr. Lyell, nearly all the species of molluscous animals in the seas of the two temperate zones are distinct, yet the whole species in one bears a strong analogy to that in the other; both differ widely from those in the tropical and arctic oceans; and, under the same latitude, species vary with the longitude. The east and west coasts of tropical America have only one shell-fish in common; and those of both differ from the shell-fish in the islands of the Pacific and the Gallipagos Archipelago, which forms a distinct region. Notwithstanding the many definite marine provinces, the same species are occasionally found in regions widely separated. A few of the shell-fish of the Gallipagos Archipelago are the same with those of the Philippine Islands, though so far apart. The east coast of America, which is poor in shell-fish, has a considerable number in common with the coasts of Europe.

The larger and more active inhabitants of the waters obey the same laws with the rest of the creation, though the provinces are in some instances very extensive. Dr. Richardson observes, that there is one vast province in the Pacific, extending 42° on each side of the equator, between the meridians including Australia, New Zealand, the Malay Archipelago, China, and Japan, in which the genera are the same; but at its extremities the Arctic and Antarctic genera are mingled with the tropical forms. Very many species of the Red Sea and eastern coast of Africa range to the Indian and China seas, those of North Australia, and all Oceanica; the continuous chains of islands being favourable to their dispersion. Few of the Pacific fish enter the Atlantic; and, from the depth and want of islands in the latter, the great bulk of species are different on its different sides. Many families are common to the colder seas in both hemispheres; but the genera are mostly different, the species always.

The British Islands lie between two great provinces of fishes—one to the south, the other to the north—from each of which we have occasional visitors. The centre of the first is on the coasts of the Spanish peninsula, extending into the Mediterranean. That on the north has its centre about the Zetland Islands; but the group peculiarly British, and found nowhere else, has its focus in the Irish Sea. It is, however, mixed with fish from the seas bounding the western shores of Central Europe, which form a distinct group.

Prince Canino has shown that there are 853 species of European fish, of which 210 live in fresh water, 643 are marine, and 60 of these go up rivers to spawn. 444 of the marine fish inhabit the Mediterranean, 216 are British, and 171 are peculiar to the Scandinavian seas; so that the Mediterranean is richest in variety of species. In it there are peculiar sharks, sword-fish, dolphins, archovies, and six species of tunny, one of the largest of edible fish, for which fisheries are established in Elba, the Straits of Messina, and the Adriatic. Four of the species are found nowhere else but in the Mediterranean. Rays of numerous species are particularly characteristic of the Mediterranean, especially the two torpedos, which have the power of giving an electric shock, and even the electric spark. The Mediterranean has two or three American species; 41 fish in common with Madeira, one in common with the Red Sea, and a very few seem to be Indian. Some of these fish must have entered the Mediterranean before it was separated from the Red Sea by the Isthmus of Suez; but geological changes have had very great influence on the distribution of fishes everywhere. Taking salt and fresh-water fish together, there are 100 species common to Italy and Britain; and although the communication with the Black Sea is so direct, there are only 27 fish common to it and the Mediterranean; but the Black Sea forms a district by itself, having its own peculiar fish; and those in the Caspian Sea differ entirely from those in every other part of the globe. The island of Madeira, solitary amid a great expanse of ocean, has many species. They amount in number to half of those in Britain; and nearly as many are common to Britain and Madeira as to that island and the Mediterranean; so that many of our fish have a wide range in the Atlantic. The Mediterranean certainly surpasses the British and Scandinavian seas in variety, though it is far inferior to either in the quantity or quality of useful fish. Cod, turbot, haddock, tusk, ling, herring, and many more, are better in northern seas than elsewhere, and several exist there only.

Whales and sharks like deep water. Different species of sharks are in all tropical and temperate seas; and, although always dangerous, they are more ferocious in some places than others, even where of the same species.

The greater number of fish used by man as food frequent shoal water. The coast of Holland, our own shores, and other parts of the North Sea, where the water is shallow, teem with a never-ending supply of excellent fish, of many kinds.

Vast numbers are gregarious and migratory. Cod arrive in the shallow parts of the coast of Norway in February, in shoals many yards deep, and so closely crowded together that the sounding-lead can hardly pass between them: 16,000,000 have been caught in one place in a few weeks. In April they return to the ocean. Herrings come in astonishing quantities in winter; and lobsters are so plentiful among the rocks in Norway, that many hundred thousands are caught every year.

The principal cod fisheries are on the banks of Newfoundland and the Dogger-bank. They, like all animals, frequent the places to which they have been accustomed. Herrings come to the same places for a series of years, and then desert them, perhaps from having exhausted the food. Pilchards, mackerel, and many others, may be mentioned among the gregarious and migratory fish.

Most lakes have fish of peculiar species, as the lake Baikal. In the North American lakes there is a thick-scaled fish, analogous to those of the early geological eras; and the gillaroo trout, which is remarkable in having a gizzard, is found in Ireland only. Forty-four fish inhabit the British lakes and rivers, and fifty those of Scandinavia, of the very best kinds. The fresh-water fish of northern climates are better than those of the southern, as salmon of various species.

Each tropical river has its own species of fish; and sea-fish, in immense quantities, frequent the estuaries of rivers everywhere. The mouth of the Mississippi is full of fish; and the quantity at the mouth of the Don, in the Sea of Azof, is prodigious.

There are some singular analogies between the inhabitants of the sea and those of the land. Many of the medusæ, two corallines, the sea-stag, and some others, sting. A cuttle-fish, at the Cape de Verde Islands, changes colour like the cameleon, assuming the tint of the ground under it. Herrings, pilchards, and many other fish, as well as sea insects, are luminous. The medusa tribe, the species of which are numerous, have the faculty of shedding light in the highest degree. In warm climates, especially, the sea seems to be on fire, and the wake of a ship is like a vivid flame. Probably fish that go below the depths to which the light of the sun penetrates are endowed with this faculty; and shoals of luminous insects have been seen at a considerable depth below the surface of the water. The glow-worm, some beetles, and fire-flies, shine with the same pale green light. But among the terrestrial inhabitants there is nothing analogous to the property of the gymnotus electricus of South America, the trembler, or silurus electricus, of the African rivers, and the different species of the torpedo of the Mediterranean, besides many others, mostly of the ray kind, in various parts of the world, which possess the faculty of giving the electric shock.

The marine mammalia form several distinct families, all of which suckle their young. Fish require air like other animals, and obtain it from the water by means of their gills; but as the whale family are not fish, they are obliged to come to the surface of the sea to breathe, which they do through nostrils.* Fat pervades every part of their body and bones, which makes them buoyant, and enables them to float without fatigue or effort; and their blood is said to be warmer than that of land animals, so that they can bear the most intense cold.

The first family of the whale tribe consists of herbivorous phocæ, as lamantins and dugons, and of predatory phocæ, as seals, and the morse or walrus, all of which have teeth and are amphibious, and some of the numerous species are found in every sea and in every latitude, but the herbivorous phocæ are mostly intertropical. Lamantins are of various sizes and kinds; the species which frequents the Antilles, the Orinoco, and Amazons, and other

* Narrative of a Whaling Voyage round the World, by F. D. Bennett, Esq.

rivers in the warmer parts of America, generally known as the sea-cow, is about 20 feet long, and has a round body, not unlike a sack of wine. It browses in herds on the herbage at the bottom and on the banks of streams, and when attacked the mother defends her young at the sacrifice of her own life, and the cub follows the dead body to the shore, so both fall an easy prey to the hunters. The dugon is not so round as the lamantin, and has a bristled snout; different species feed in flocks on the weeds at the bottom of shallow parts of the Indian Ocean, the Indian Archipelago, the coasts of Africa, New Holland, and the Red Sea, and never enter fresh water. They are so harmless and tame that they allow themselves to be handled, and they sit upright when they suckle their young, which has given rise to the fable of the Mermaid. This animal sacrifices her life for her young like the lamantin, and is the type of maternal affection among the Malays. The manatus septentrionalis is the only herbivorous seal that is found in the Arctic Ocean; it frequents the Spitzbergen seas, but is very rare.

The favourite haunts of the predatory seals are the polar oceans and desert islands in high latitudes, where they bask in hundreds on the sunny shores during the brief summer of these inhospitable regions, and become an easy prey to man, who has nearly extirpated the race in many places. Six or seven species of seals are natives of Iceland, and two or three of Greenland. The common seal is six or seven feet long, with a face like that of a dog, and a large intelligent eye. It is easily tamed, and in the Orkney Islands it is so much domesticated that it follows its master, and helps him to catch fish. This seal migrates in herds from Greenland twice in the year, and returns again to its former haunts; they probably come to the coasts of Europe and the British Islands at the time of their migrations, but the phoca vetulina is a constant inhabitant of our shores. Some of the seal tribe have a very wide range, as the fur species, arctocephalus ursinus, of the Falkland Islands, which at one time frequented the southern coasts of New Holland in multitudes, but they and three other species have now become scarce from the indiscriminate slaughter of old and young. Sir James Ross found some of the islands in the Antarctic seas overrun with the sea elephant, moremga elephantina, and they captured a new species of seal without ears. The walrus or morse, a grim-looking creature, with tusks two feet long bent downwards, and its face covered with transparent bristles, has a body like that of a seal, 20 feet long, with a coat of short grey or yellow hair. It sleeps on the floating ice, feeds on sea-weeds and marine animals, and never leaves the Arctic seas.

The second family of marine mammalia consists of spouting whales of predatory habits; they live on fish, and consequently have teeth, such as porpoises, dolphins of many kinds, and spermaceti whales or cachelots; these have spouting nostrils in the upper part of the head. The common porpoise is seen spouting and tumbling on the surface of all the seas of Europe, shoals of them go in pursuit of herrings and mackerel, and even swim up the rivers in chase of salmon. They have more the form of fish than the seal tribe, and have a dorsal fin. The common dolphins, so remarkable for voracity and for the swiftness of their motions, which is owing to the symmetry of their form and the width of their tail, are seen in almost every latitude and sea, but probably of different species. The white dolphin, eaten by the Icelanders, is 18 feet long, and migrates from the Atlantic to Greenland in the end of November. The grampus, delphinus orca, possibly the same with the killer of the South Sea whalers, is a fierce voracious fish, often 20 feet long, which roams in numerous shoals, preying upon the larger fish, and even attacking the whale. The grind or black dolphin has been known to run ashore in hundreds in the bays of Feroe, Orkney, and Zetland. This seems to be the same or nearly allied to the black fish, which was met with in vast numbers by Sir James Ross in the Antarctic seas: they had so little fear, that they darted below the ship on one side and came up at the other. The right white porpoise, delphinus peronii, of the southern whalers, is a rare and elegant species of dolphin which chiefly inhabits the high southern latitudes, but has been seen at the equator in the Pacific. They are about six feet long, the hinder part of the head, the back, and the flukes of their tail are black, and all the rest of the purest white. The narwhal or sea unicorn, monodon monoceros, has no teeth, but a tusk of fine ivory wreathed with a spiral grove extending eight or ten feet straight from the head; in general there is only one tusk, but there are always the rudiments of another, and occasionally both grow to an equal length. The old narwhals are white with blackish spots, the young are dark-coloured. This singular creature, which is about 16 feet long without the tusk, swims with great swiftness. Mr. Scoresby has seen 15 or 20 at a time playing round his ship in the Arctic seas, and crossing their long tusks in all directions as if they were fencing; they are found in all parts of the Northern Ocean.

The spermaceti whale, the cachelot or physeter macrocephalus, belonging to the family of the predaceous spouters, is one of the most formidable inhabitants of the deep. Its average size is 60 feet long and 40 feet in circumference; its head, equal to a third of its length, is extremely thick and blunt in front, with a throat wide enough to swallow a man. The proportionally small swimming paws or pectoral fins are at a short distance behind the head, and the tail, which is a horizontal triangle six or seven feet long, and 19 feet wide, with a notch between the flukes, is the chief organ of progressive motion and defence. It has a hump of fat on its back, is of a dark colour, but with a very smooth clean skin. These sperm whales have one nostril on the top of their head, through which they throw in breathing a continued succession of jets like smoke, at intervals of 15 or 20 minutes, after which they toss their tails high in the air and go head foremost to vast depths, where they remain for an hour or an hour and a half, and then return again to the surface to breathe. The jet or spout is from six to eight feet high, and consists of air expired by the whale, condensed vapour, and particles of water. This whale has sperm-oil and spermaceti in every part of its body, but the latter is chiefly in a vast reservoir in its head, which makes it very buoyant, and ambergris is sometimes found in the inside of the body, supposed to be from disease. These huge monsters, occasionally 75 feet long, go in great herds of 500 or 600, or schools, as the whalers call them. Females with their young, and two or three old males, generally form one company, and the young males another, while the old males feed and hunt singly. The sperm whales swim gracefully and equally, with their head above the water; but when a troop of them play on the surface of the water, some of these uncouth and gigantic creatures leap with the agility of a salmon several feet into the air, and fall down again heavily with a tremendous crash and noise like a cannon, driving the water up in lofty columns capped with foam. The fishery of the sperm whale is attended with great danger; not only the wounded animal, but its companions who come to its aid, sometimes fight desperately, killing the whalers and tossing them into the air with a sweep of their tremendous tails, or biting a boat in two. In 1820 the American whaler Essex was wrecked in the Pacific by a sperm whale; it first gave the ship so severe a blow that it broke off part of the keel, then retreating to a distance, it rushed furiously, and with its enormous head beat in a portion of the planks, and the people had just time to save themselves in the boats when the vessel filled. They often lie and listen when suspicious of mischief. No part of the aqueous globe, except the Arctic seas, is free from their visits; they have been seen in the Mediterranean, the British Channel, and even the Thames, but their chief resort is the deepest parts of the warmer seas within or near the tropics, and in the Antarctic Ocean, where they feed on floating shell-fish and the sepia or cuttle-fish.

The third and last family of marine mammalia are whalebone whales, such as the Greenland whales and rorquals. Instead of teeth, the jaws of these animals are furnished with plates and filaments of whalebone, which are moveable, and are adapted to retain, as in a net, the medusæ and other small marine animals that are the food of these colossal inhabitants of the deep. The common Greenland species, balæna mysticetus, was formerly much more numerous, but it is now chiefly confined to the very high northern latitudes; however should it be the same with the whale found in such multitudes in shallow water on the coasts of the Pacific and in the Antarctic Ocean by Sir James Ross, it must have a very wide range, but it is more probable that each pole has its own species. The Greenland whale is from 65 to 70 feet long, but they are so much persecuted that they probably never live long enough to come to their full size. The head is very large, but the opening of the throat is so narrow that it can only swallow small animals. It has no dorsal fin: the swimming paws are about nine feet long, and the flat tail is half-moon shaped and notched in the middle. It has two spouts or nostrils, through which it throws jets like puffs of smoke some yards high. It only remains two or three minutes on the surface to breathe, and then goes under water for five or six. The back and tail are velvet-black, shaded in some places into grey, the rest is white: some are piebald. The capture of this whale is often attended with much cruelty, from their affection for their young; indeed the custom of killing the calf in order to capture the mother has ruined the fishery in several places, especially in New Zealand, where there were eight species of whales in vast abundance.

Rorquals are also whalebone whales, differing somewhat in form from the common whale. One species is from 80 to 100 feet long, the largest of marine animals. The bottle-nosed whale, a smaller species, was exceedingly numerous in the Arctic seas; in the year 1809, 1100 were stranded in Huelfiord in Iceland. This whale travels to lower latitudes in pursuit of herrings and other fish. It had been caught on the coast of Norway as early as the year 890, and probably long before. The first northern navigators were not attracted by the whale as an object of commerce, but stumbled upon it in their search for a north-west passage to the Pacific.

The hump-backed whale, balæna gibbosa, a rorqual 30 or 40 feet long, is met with in small herds in the intertropical and southern regions of the Pacific and Atlantic; it is seldom molested by the whalers, and is very dangerous for boats, from the habit it has of leaping and rising suddenly to the surface. None of the senses of the whale tribe are very acute; the whalebone whales alone have the sense of smelling, and although the sperm whale is immediately aware of a companion being harpooned at a very great distance, they do not hear well in air, and none have voice.*

* Captain Scoresby's Arctic Voyages.

CHAPTER XXVIII.

DISTRIBUTION OF REPTILES—FROGS AND TOADS—SNAKES, SAURIANS, AND TORTOISES.

REPTILES, more than any other class of animals, show the partial distribution of animated beings, because, being unable to travel to any great distance, they have remained in the places wherein they were originally stationed; and as they inhabit deserts, forests, and uncultivated ground, they have not been disturbed by man, who has only destroyed some individuals, but has not dimimished the number of species, which is probably the same as ever it was. Few of the mammalia hybernate, or fall into a torpid state in winter, except the bear, marmot, bats, and some others. Their fat supplies the carbon consumed by the oxygen during their feeble and imperceptible respiration, and is wasted by the time the warm weather returns, which rouses them from their lethargy, thin and extenuated. But reptiles, being colder blooded, bury themselves in the ground, and hybernate during the winter in cold and temperate climates. In hot countries, they fall into a state of torpor during the dry season, so that they have no occasion to wander either on account of temperature or want of sustenance; and the few that do migrate in quest of food, always return to their old haunts. As the blood of reptiles receives only a small part of the oxygen they inhale, little heat and strength are generated; consequently they are cold-blooded, and for the most part sluggish in their motions, which, however, are more varied than in quadrupeds; but as some reptiles, as tortoises and lizards, breathe more frequently than others, there are consequently great differences in their energy and sensibility. There are four distinct classes of reptiles—toads and frogs, serpents, lizards, and tortoises. These partake of both terrestrial and aquatic forms, and many are amphibious: they all increase in numbers towards the equator, and few live in cold climates.

The toad and frog class approaches nearest to the nature of fishes, and forms a link between land and water animals. As tadpoles they have tails and no feet, but when full-grown, they generally acquire feet and lose their tails. Besides, in that early stage they are aquatic, and breathe by gills, like fishes; but in a state of maturity they breathe by lungs, like quadrupeds, though some of the families always retain their gills and tails, and some never acquire feet. These animals have the power of retarding and accelerating their respiration without stopping the circulation of their blood, so that they can resist heat and cold to a certain degree—a power most remarkable in the salamander, which forms part of this class, so varied in appearance and nature. Some, as toads and frogs, imbibe a quantity of water, which is evaporated through their skin more or less quickly. This keeps them at the temperature of the medium they live in, and the air they inhale through the skin is as necessary to their existence as that which they breathe.

The group of toads and frogs consists of four families, which have four feet, but neither necks nor tails ; namely, frogs, hylas or rainettes, toads, and pipæ. Frogs, which are amphibious, have no nails on their toes, and their hind legs are webbed, consequently fitted for swimming, which they do by leaps. There are 16 genera, and above 50 species, so that they are more numerous and more varied than any other reptile. Of the hyla, rainette, or tree-frog, there are 60 species, all of the most vivid and brilliant tints, and several colours are frequently united on the same animal. They spend most of their lives on high trees, and their webbed feet have little cushions at the points of their toes, by means of which they can squeeze out the air from under their feet, and, by the pressure of the atmosphere, they adhere firmly to the under side of the smoothest leaf, exactly on the same principle by which flies walk on the ceiling of a room. The bufo, or toad, is the ugliest of the race; many are hideous, with swollen bodies and obtuse toes. They do not go into water, but frequent marshy, damp places, and only crawl, whereas the frog and hyla leap. They are much fewer than either of the other two families; only thirty species are known. The pipæ are also toads of a still more disgusting form, and are distinguished from their congeners by having no tongue. There are only two species known. All these reptiles have voices, which are exceedingly varied; they croak in concert, following a leader, and when he is tired another takes his place. One of the North American frogs croaks in bands; one band begins, another answers, and a third replies, till the noise is heard at a great distance; a pause then takes place, after which the croaking is renewed. Mr. Darwin mentions a little musical hyla at Rio de Janeiro, which croaks a kind of harmony in different notes.

Toads and frogs are found in almost all parts of the earth, though very unequally and partially distributed. America has more than all the other countries taken together, and Europe the fewest. Six species of frogs, one rainette, and two toads, are European; and all, except four of the frogs, are also found in Asia and Africa.

The law of circumscribed distribution is strongly marked in Asia; for of ten species of frogs peculiar to that continent, three only are in the mainland, two are confined to Japan; and of the five that are Javanese, one is also common to Amboina, and the other four to Bengal. The eight species of rainettes, or tree-frogs, are still more limited in their domicile; five of them are in Java only, and one in Japan. There are nine species of toad peculiar to Asia.

None of these reptiles exist in the Galapagos Archipelago, nor in any of the innumerable islands in Oceanica, and there are very few in Australia, but all peculiar. In Africa there are eight species of frogs, two or three of rainettes, and two of toads. One of the two species of pipas, more horrid in appearance than any toad, is very common at the Cape of Good Hope, and there only.

The great extent of marshes, rivers, and forests, together with the heat of the climate, make America the very home of reptiles of this kind, and there they grow to a greater size than anywhere else: 23 species of frog, 27 species of tree-frog or rainette, and 21 of toads, are indigenous in that continent, not one of which is the same with any of those in the old world; and most of those in South America are different from those in the northern part of the continent, though they are sometimes replaced by analogous kinds. All these reptiles have abodes, with fixed demarcations, often of small extent. The pipa, or toad of Surinam, is the most horrid of the tribe; the bufo aguà, of Brazil, 10 or 12 inches long, and the rana pipiens, of Carolina, are the largest.

The second family of this class of reptiles have tails and feet, as the salamanders, which are very like lizards, with a long round tail and four feet. Some are terrestrial, and some are aquatic; the latter are known as tritons. Both are in Europe, but the greater number are American; and the sirens are peculiar to the marshes and rice-grounds of Carolina. They are very like eels with two feet. The proteus anguinus is similar, but it has four little feet and a flat tail, and has been found nowhere but in the dark subterraneous caverns in Carniola.

The third family of this class of reptiles is the

cæciliæ, of which there are only eight species, all inhabitants of the warm parts of Asia, Africa, and America. They have a cylindrical body, without feet or neck, and move exactly as the serpent, so they seem to form the link between these reptiles and the class of frogs and toads.

There are serpents in all hot and temperate countries, but they abound most in intertropical regions; and wherever snakes exist, there also are some of the venomous kinds, but they are fewer specifically and individually than is generally supposed. Of 263 species, only 57 are venomous, or about one in five, although that proportion is not everywhere the same. In sterile, open countries, the proportion of venomous snakes is greater than in those that are covered with vegetation. Thus, in New Holland, seven out of ten species are poisonous; and in Africa, one of every two or three individuals is noxious. In general, however, the number of harmless individuals is 20 times as great as the number of the poisonous.

The three great families of venomous serpents are the colubriform or adder-shaped snakes, sea-serpents, and the triangular-headed snakes.

The adder-formed snakes are divided into three genera, the elaps tribe, which are slender like a cord, with a small head and of brilliant colours. There are four species in South America, of which two are confined to Guiana, and one to Surinam, while the other is found everywhere from Brazil to Carolina. There is only one in Africa, three in

Australia, and the rest are in limited districts in tropical Asia, especially in Sumatra and Java; and an entire genus is found only in India, and the islands of Ceylon and Java. The hooded snakes are the best known of this family, especially the spectacled or dancing snake of the Indian jugglers, which is common everywhere from Malabar to Sumatra, and two other species are only in Sumatra and Java. The three or four African species are chiefly at the Cape of Good Hope and on the Gold Coast; but the most celebrated is that generally known as the Egyptian asp, which has been tamed by magicians of ancient and modern times, and is frequently figured in Egyptian monuments; it derives some of its celebrity from Cleopatra's death. Two of the family inhabit New Holland, one of which is spectacled, but of a different species from that in India.

All the seven species of sea-snakes are very venomous, and more ferocious than any other. They frequent the Indian Ocean in shoals from Malabar to the Philippine Islands, but chiefly the Bay of Bengal; they never enter fresh water, nor do they ever land.

The third venomous family consists of the triangular-headed serpents, rattle-snakes, and vipers. The first are of a hideous aspect,—a large head, broad at the base like a heart, a wide mouth, with their hooked poisonous fangs strongly developed. They quietly watch their prey till it is within reach, then dart upon it, and inflict the deadly wound in a moment. There are four species of these formidable snakes in the intertropical parts of South America, and in the warmer parts of North America. One species in the old world is to be met with everywhere, from Ceylon to the Philippine Islands; one is a native in Sumatra, Timor, and Celebes; the rest are narrowly limited in their abode; two are confined to Java alone. Ceylon, Sumatra, Japan, and Tartary, have each a species of these serpents peculiar to itself.

The rattle-snakes are all American—two in the warm districts of North America, and two in the intertropical parts of South America. One of the latter, however, has a hard horn at the end of its tail, instead of a rattle, and sometimes grows to the length of 10 feet, being the longest of the venomous snakes.

Vipers come farther north than any other of the noxious tribe: two are Asiatic, though one is also common to Africa, which, however, has four peculiar to itself: and the only venomous serpents in Europe are three species of viper, one of which is also spread over the neighbouring parts of Asia and Africa. The common viper inhabits all central Europe and temperate Asia, even to Lake Baikal, in the Altaï Mountains: it is also found in England and Sweden, but it does not go farther west than the Seine, nor does it pass the Alps. One which frequents dry soils, in the south-east of Europe, is in Styria, Greece, Dalmatia, and Sicily; and the aspic viper, which lives on rocky ground, inhabits France between the Seine and the Pyrenees, Switzerland, Italy, and Sicily.

There are six families of harmless serpents, consisting of numerous species. Four of the families are terrestrial, their species are very limited in their domicile, the greater number being confined to some of the islands of the Indian Archipelago, Ceylon, or to circumscribed districts in tropical Asia, Africa, and America. Nine or ten species are European, some of which are also found in Asia and Africa.

Tree-serpents of various genera and numerous species live only in the great tropical forests of Asia and America, especially in the latter. They are long and slender, the head for the most part ending in a sharp point, and generally green, though there are some of brighter colours; many of these serpents are fierce, though not venomous; some feed on birds, which they watch hanging by the tail from a bough.

In all temperate and warm countries abounding in lakes and rivers, fresh-water snakes are numerous; some live in the water, but they mostly inhabit the banks near it; they are excellent swimmers, and may be seen crossing lakes in shoals. America is particularly rich in them; there are several in Europe and Asia, but they are rare in Africa, and none have been yet discovered in Australia.

The boa is peculiarly American, though some smaller in size and differing in species are found in Asia. The boa constrictor, generally from 9 to 12 feet long, lives in the great tropical forests of South America, where it watches its prey hanging from the boughs of trees. Two of smaller size have similar habits, and two are aquatic, one of which is sometimes 20 feet long, and another 6 feet; the latter inhabits the rivers from the Amazons to Surinam.

Pythons are the largest snakes of the eastern world; one species, which sometimes grows to the length of 20 feet, is spread from the western coast of Africa, throughout intertropical Asia to Java and China. Another, only 14 feet long, is confined to Malacca and some of the Sunda Islands. Two others are found only in the islands of Timor and Saparua, and one in New Holland. There are only two species of acrochordi, which, like boas and pythons, twist themselves round their victims and crush them to death: one aquatic, peculiar to Java; the other is a land snake, found everywhere through India to New Guinea.

The West Indian Islands have the snakes of North and South America, and some peculiar; the snakes of central America are little known.

Saurians have representatives in every warm and temperate climate. The crocodile, from its size and ferocity, claims the first place. There are three genera of this family, all amphibious, living in rivers: the crocodile, common to the old and new continents; the caiman, or alligator, peculiar to America; and the gavial, which comes nearer to the form of the fossil crocodile than any other, is limited to the Ganges and other great rivers of India. The various species of crocodiles are confined to local habitations: three are Asiatic; two African, one of which is only in Sierra Leone; two are peculiar to Madagascar; and in America there are two species of crocodiles and five of alligators. The American crocodiles inhabit the estuaries of great rivers; the alligator never leaves fresh water.

The alligators of the Mississippi, and of the rivers and marshes of Carolina, are more ferocious than those of South America, attacking men and animals; they only prey in the night; while in the water, they cannot swallow their food, but they drown the animal they have caught, hide it under water till it is putrid, and then bring it to land to eat it. Locality has considerable influence on the nature and habits of these animals; in one spot they are very dangerous, while in another, at no great distance, they are cowardly. Alligators are rarely more than 15 feet long, and are seen in great companies basking on the banks of rivers: their cry is like the roar of a bull; in a storm they bellow loudly, and are said to be much afraid of some of the whale family that ascend the great American rivers. The female watches her eggs and her young for months, never losing sight of them; but the male devours many of them when they go into the water. All animals of this class are covered with scales, those of the crocodile family are hard and impenetrable.

Lizards are chiefly distinguished from crocodiles by having a long, thin, forked tongue like that of the viper; by their rapid motions, smaller size; and by some peculiarities of form.

The monitors, which are entirely confined to the old continent, have the tail compressed laterally, which enables them to swim rapidly; and they are furnished with strong sharp teeth. Many species inhabit Africa and India, especially the Indian Archipelago: the terrestrial crocodile of Herodotus is common on the deserts round Egypt; and an aquatic species in the Nile, which devours the crocodile's eggs, is often represented on the ancient Egyptian monuments.

Another group of the monitor family is peculiarly American; some of the species inhabiting the marshes in Guiana are six feet long.

Lizards are very common; more than eight or nine species are European: and the iguanians, which differ from them only in the form of the tongue, are so numerous in genera and species, that it would be in vain to attempt to follow all their ramifications, which are, nevertheless, distributed according to the same laws with other creatures: but the dragons, only found in India, are too singular to be passed over. The dragon is in fact a lizard with wings of skin, which are spread along its sides and attached to its fore and hind feet, like those of the bat, and though they do not enable it to fly, they act like a parachute when the animal leaps from bough to bough in pursuit of insects. Nocturnal lizards of many species inhabit the hot countries of both continents; they are not unlike salamanders, but they have sharp claws, which they can draw in and conceal like those of a cat, and seize their prey. One of this species climbs on walls in all the countries round the Mediterranean. Chameleons are to be seen under every bush in North Africa; and different species inhabit different districts and islands in Asia; the only one that is European lives in Spain; it is also common to North Africa.

The anolis, which lives on trees, replaces the chameleon in the hot regions of South America, and in the Antilles, having the property common to chameleons of changing its colour, but it is a more nimble and beautiful animal. In New Holland, where every thing is anomalous, there is a lizard with a leaf-shaped tail.

Skinks are exactly like serpents, with four very short feet and sharp nails on their claws, which burrow in the sands of Africa and Arabia: there is a species of gigantic black and yellow skinks in New Holland, and those in the islands of the Indian Archipelago are green, with blue tails.

Two anomalous saurians of the genus amblyrhinchus were discovered by Mr. Darwin in the Galipagos Archipelago. One found only in the central islands is terrestrial, and in many places it has undermined the ground with its burrows; the other is the only lizard known that lives on sea-weed, and inhabits the sea; it is about four feet long, and hideously ugly, with feet partially webbed and a tail compressed laterally. It basks on the beach, and in its marine habits and food it resembles, on a small scale, the huge monsters of a former creation.

Tortoises are covered with a shell or buckler, but their head, legs, and tail are free, covered with a wrinkled skin, and the animal can draw them into the shell when alarmed. The head is sometimes defended by a regular shield, and the jaws, instead of teeth, have a horny case. The upper buckler is rounded, and formed of eight pairs of plates symmetrically disposed, and often very beautiful; the under shell is flat, and consists of four pair of bones and one in the centre. One family of tortoises is terrestrial, two others are amphibious, one of which lives in fresh water, the other in tropical and warm seas.

There are more land tortoises in Africa than in all the rest of the world, both specifically and individually. There are several European species, of which the Greek tortoise, common in all the countries round the Mediterranean, is the largest, being about a foot long; it lives on insects and vegetables, and burrows in the ground in winter. Some of the East Indian species are enormously large, above three feet long, and remarkable for the beautiful distribution of their colours; some species are peculiar to Brazil, one to Demarara, and one to North America.

There are two families of the fresh-water tortoises that live in ponds and ditches. The emys is very numerous in America; there are 15 species peculiar to the northern part of the continent, and four to the southern: only one has been found in Africa, two in Europe, and eight in Asia. South

America is the country of the chelydæ, which feed only when in the water: there are none in North America, five in Africa, and one in Australia.

The trionyx, or fresh-water turtle, lives in the great rivers and lakes in warm countries; there are two species peculiar to North America, they are very large, eat birds, reptiles, and young crocodiles, and often are a prey to old ones. One is peculiar to the Nile, one to the Euphrates, but the Ganges is their principal abode; there are four species which are constantly seen eating the bodies of the natives that are thrown into this sacred stream; one of these turtles often weighs 240 pounds. The starred trionyx is in the rivers of Java only, and another kind is common also to the rivers of Borneo and Sumatra.

The cheloniadæ, or sea turtle, live in the seas of the torrid and temperate zones, to the 50th parallel of latitude, some eating algæ, and others molluscas and radiated fish. Different species are found in different parts of the ocean. The green turtle, of which there are many varieties, inhabits the inter-tropical Atlantic; they are seen in shoals eating sea-weed at the bottom of the water along the coasts, but they come to the mouth of rivers to lay their eggs in the sand. This turtle is often six or seven feet long, and weighs 600 or 700 pounds; it is much esteemed for food, but the shell is of no value. There are two species in the Mediterranean, which are only valued for the oil.

With respect to the whole class of reptiles it

may be observed, that not one species is common to the old and new world, and few are common to North and South America; those in New Holland are altogether peculiar; and, with the exception of the Marianne Islands, there are neither toads, frogs, nor snakes in any of the islands of Oceanica, though the Indian Archipelago abounds in them.

Five species of reptiles only had reached Ireland before its separation from England, a lizard, a frog, a toad, and two tritons.

CHAPTER XXIX.

DISTRIBUTION OF BIRDS IN THE ARCTIC REGIONS — IN EUROPE, ASIA, AFRICA, AMERICA, AND THE ANTARCTIC REGIONS.

There is great similarity in the birds of the northern parts of the old and new continents, and many are identical. Towards the south, the forms differ more and more, till in the tropical and south temperate zones of Asia, Africa, and America they become entirely different, whole families and genera often being stationary within very narrow limits. Some birds, however, are almost universal, especially birds of prey, waders, and sea-fowl.

The bald buzzard is in every country, from Europe to Australia; the Chinese gosshawk inhabits the American continent, and every station between China and the west coast of Europe: the peregrine falcon lives in Europe, America, and Australia; the common and purple herons are indigenous in the old continent and the new; and the flamingo fishes in almost every tropical river. Many of the sea-fowl also are widely spread: the wagel-gull is at home in the northern and southern oceans, and on the coasts of Australia. Captain Beechey's ship was accompanied by pintadoes during a voyage of 5000 miles in the Pacific; and even the common house-sparrow is as much at home in the villages in Bengal as it is in Britain. Many more instances might be given, but they do not interfere with the general law of special distribution.

Birds migrate to very great distances in search of food, passing the winter in one country and the summer in another, many breeding in both. In cold climates insects die or hybernate during winter; between the tropics, they either perish or sleep in the dry season: so that, in both cases, insect-eating birds are compelled to migrate. When the ground is covered with snow, the want of corn and seeds forces those kinds whose food is vegetable to seek it elsewhere; and in tropical countries the annual inundations of the rivers regulate the migrations of birds that feed on fish.

Some migrate singly, some in groups, others in flocks of thousands; and, in most instances, the old and the young birds go separately. Those that fly in company generally have a leader, and such as fly in smaller numbers observe a certain order. Wild swans fly in the form of a wedge, wild geese in a line. Some birds are silent in their flight, others utter constant cries, especially those that migrate during night, to keep the flock together, as herons, goat-suckers, and rails.

Birds of passage in confinement show the most insurmountable disquietude when the time of migration draws near. The Canadian duck rushes impetuously to the north at the usual period of summer flight. Redbreasts, goldfinches, and oriols, brought from

Canada to the United States when young, dart northwards, as if guided by the compass, as soon as they are set at liberty. Birds return to the same place year after year. Storks and swallows take possession of their former nests, and the times of their departure are exact even to a day. Various European birds spend the winter in Asia and Africa; while many natives of these countries come to Central Europe in summer.

The birds of passage in America are more numerous, both in species and individually, than in any other country. Ducks, geese, and pigeons migrate in myriads from the severity of the northern winters; and when there is a failure of grain in the south, different families of birds go to the north. The Virginian partridge crosses the Delaware and goes to Pennsylvania, when grain is scarce in New Jersey; but it is so heavy on the wing, that many fall into the river, and end the journey by swimming. The same thing happens to the wild turkey, which is caught in hundreds as it arrives wet on the banks of the Ohio, Missouri, and Mississippi. These birds are not fitted for long flight by their structure, because their bones have fewer of those air-cells which give buoyancy to the feathered tribes. The number of air-cells is greatest in birds that have to sustain a continued and rapid flight; probably the extremes are to be met with in the swift and the ostrich—the one ever on the wing, the other never. The strength of the ostrich is in the muscles of its legs; while the muscles on the breast of the swift weigh more than all the rest of the body: hence it flies at the rate of 100 miles an hour easily. The wild duck and the wild pigeon fly between 400 and 500 miles in a day. The stork and some other migratory birds do not halt till the end of their journey. Many sea-fowl are never seen to rest; and all the eagles, vultures, and hawks are birds of strong flight, and capable of sustaining themselves at heights beyond the reach of less buoyant creatures.

DISTRIBUTION OF ARCTIC AND EUROPEAN BIRDS.

The birds of Europe and North America are better known than those of any part of the globe. New species are constantly discovered in Asia, Africa, and South America; and extensive regions in the East are yet unexplored: however, about 6000 have already been described.

There are 503 species of birds in Europe, many of which are distributed over Asia and Africa, without any apparent variation; and 100 of our European species are also in North America. Of these, 39 are land-birds, 28 waders, and 62 water-fowl; among which are most of the marine birds of northern Europe, which, like all sea-fowl, have a wider range.

More than three-fourths of the species, and a much larger proportion of individuals, of the birds of Greenland, Iceland, and Feroe, are more or less aquatic, and many of the remainder are only occasional visitors. Of the few small birds, the greater number are British; but many that reside constantly in Britain are migratory in Iceland and Feroe, and all the small birds leave Greenland in winter. The aquila albicilla, or cinereous eagle, is the largest bird of these northern islands; it feeds on salmon and trout, and builds its nest on the boldest crags. The jer-falcon, or falco Icelandicus, though native, is rare even in Iceland. The snowy owl lives near the glaciers in the interior of Greenland, and is sometimes seen in Orkney. Particular kinds of grouse are peculiar to high latitudes, as the ptarmigan and the white grouse. The columba ænas lives on all the rocky coasts of Europe, and it is also an American bird. The crow family are inhabitants of every part of the globe. The common crow is universal; the carrion-crow and jackdaw are all over Europe and North America. The magpie is everywhere in Europe. The jay, one of the most beautiful birds of its tribe, is found in Europe, North America, and China. The raven is everywhere, from Greenland to the Cape of Good Hope, and from Hudson's Bay to Mexico; it is capable of enduring the extremes of heat and cold, and is larger, stronger, and more ravenous in the Arctic islands than anywhere else. It destroys sheep and lambs, drives the eider-ducks from their nests to take their eggs or young, and they unite in flocks to chase intruding birds from their abode.

Waders are more numerous than land-birds in the Arctic regions. The snipe is a resident; the golden plover is in Feroe only; and the oyster-catcher remains all the year in Iceland: it makes its nest near streams, and wages war with the crow tribe. The heron, curlew, plover, and most of the other waders, emigrate.

Web-footed birds, being clothed with down and oily feathers, are best able to resist the cold of a polar climate. The cygnus musicus, or whistling swan, is the largest migratory bird of Europe or America. It is five feet long from the tip of the bill to the end of the tail, and eight feet from tip to tip of the wings: its plumage is pure white, tinged orange or yellow on the head. Some of them winter in Iceland; and in the long Arctic night their song is heard, as they pass in flocks: it is like the notes of a violin. Various species of the duck tribe live in the far north, in prodigious multitudes. The mallard, supposed to be the origin of our tame duck, is everywhere in the Arctic lands. There are two species of eider-duck: the king duck, or somateria spectabilis, one of these, is widely dispersed over the islands and coasts of the North Atlantic; it lives in the open sea in winter, and resorts to the coast when the grass begins to grow. The duck makes her nest of sea-weed, lined with down from her breast. The islanders take the eggs and down twice in the season; but they do not kill the old birds, because the down of a dead duck is of no value, having lost its elasticity. The third time the drake repairs the nest with down from his breast: the birds are allowed to hatch their brood; and, as soon as the young can feed themselves, they are taken out to sea by the duck. They attain maturity in four years, and then measure two feet from tip to tip of the wing. The same couple has been known to frequent a nest 20 years, and the Icelanders think the eider-duck lives to 100.

The cormorant is universal in the northern seas, and, though living on fish, it is eaten by the natives. It sits singly, or sometimes in flocks, on the rocks, watching the fish with its keen eye: it plunges after them, and pursues them for three or four minutes under water. Auks are very numerous, especially the razor-billed auk, or penguin; but the great auk, which is incapable of flight with its little wings, is now extinct in the Arctic islands. The tern, or sea-swallow, is seen everywhere in these seas, skimming along the surface of the water, catching molluscas and small fish. Gulls of many species, and in countless numbers, are inhabitants of the Arctic and Antarctic regions. No birds are more widely dispersed. They are at home, and brave the storm, in every latitude and in every sea; but those in the north are said to be larger and more numerous than elsewhere. There are nine or ten species in the Arctic regions, and the most numerous of these probably are the kittywakes, the young of which cover the rocks in Iceland, packed so close together, that 50 are killed at a shot.

The skua is one of the boldest and most rapacious of birds, forming a link between gulls and birds of prey. It lives by robbing other birds, and is so audacious that it forces the gulls to disgorge the fish they have swallowed, and has been seen to kill a puffin at a single blow. Its head-quarters are in Feroe, Zetland, and the Hebrides, where it hatches its brood, and attacks men or animals if they come near them.

Several kinds of petrels inhabit the Arctic islands. They take their name from the faculty they have of walking on the water, which they do by aid of their wings. The stormy petrel, the most widely diffused, is about the size of a lark, and nearly of the colour; their flight is rapid; they shelter themselves from the storm in the hollow of a wave, and go to land only at the breeding season.

It is observed that all birds living on islands fly against the wind when they go to sea, so as to have a fair wind when they return home tired. The direction of the prevailing winds, consequently, has great influence on the choice of their abode: for example, the 25 bird-rocks, or Vogel-berg, in Feroe, face the west or north-west; and no bird frequents the cliffs facing the east, though the situation is to all appearance equally good; a preference accounted for by the prevalence of westerly wind in these latitudes.

Most marine birds are gregarious. They build their nests on the same rock, and live in society. Of this a curious instance occurs on the rocks in question. The Vogel-berg lies in a frightful chasm among the cliffs of Westmannshaven in Feroe. The chasm is encompassed by rocks 1000 feet high, and myriads of sea-fowl cluster round the top of the crags; but different kinds have separate habitations; and no race or individual leaves his own quarters, or ventures to intrude upon his neighbours.

Upon some low rocks, scarcely rising above the surface of the water, sits the glossy cormorant; the predatory skuas, on a higher shelf, are anxiously regarded by myriads of kittywakes on nests in crowded rows along the shelving rock above, with nothing visible but the heads of the mothers almost touching one another; the auks and guillemots are seated a stage higher on the narrow shelves, in order as on a parade, with their white breasts facing the sea, and in absolute contact. The puffins form the summit of this feathered pyramid, perched on the highest station, and scarcely discernible from its height, if they did not betray themselves by flying backwards and forwards. Some of these tribes have a watch posted to look out for their safety; and such confidence has the flock in his vigilance, that if he is taken the rest are easily caught. When the whole take flight, the ear is stunned by their discordant screams.*

The greater part of the marine birds of the Arctic seas are inhabitants also of the northern coasts of the continent of Europe and of the British Islands.

No part of Europe is richer in birds than Britain, both in species and numbers of individuals; and the larger game is so abundant, that no one thinks of

* Trevelyan's Travels in Iceland and the Feroe Islands.

eating nightingales and redbreasts. Of the 503 species of European birds, 277 are native in our islands. The common grouse, the yellow and pied wagtails, and the English starling, are found nowhere else. Most of the British birds came from Germany before the separation of our island from the continent, and many of short flight never reached Ireland. The ptarmigans and capercaily came from Norway.

There are five European vultures: the lammergeyer of the Alps and Pyrenees, the largest of these, builds its nest in the most inaccessible parts of the mountains, and is seldom seen; it lives also in the mountains of Abyssinia and on the Mongolian steppes. Ten eagles are European; one is peculiar to Sardinia; and several of them are common in America: the golden eagle is one; that beautiful bird, which once gave a characteristic wildness to our Scotch mountains, and the distinguishing feather to the bonnet of our chieftains, is now nearly extirpated. The osprey or fishing eagle is equally an inhabitant of Europe and America, and so are some of our numerous hawks; among others the jer or gentil falcon has been so much destroyed, that it is now rare even in Iceland, its native place: there are still a few in Scotland, and several are caught in their migratory flight over the Low Countries and reclaimed by the expert falconers for the now nearly obsolete sport of falconry.

The owl tribe is numerous, and many of them are very handsome. The bubo maximus, the great owl, the largest of nocturnal birds, inhabits the forests of middle and southern Europe; it is rare in France and England, though not uncommon in Ireland and Orkney: in Italy a small owl is tamed and used as a decoy.

Owls, eagles, and hawks have representatives in every country, but of different species. The two species of European goatsuckers migrate to Africa in winter; their peculiar cry may be heard on a moonlight night when a large flock takes wing for the journey. Several of our swallows go to Africa: both our kingfishers are African, and only visit us in summer; one, the alcyone ispida, is a native of Lower Egypt and the Red Sea. Some of the seven species of European creeping birds, or certhias, creep on the trunks and branches of trees in search of insects; others pursue their prey clinging to the face of rocks and walls, supported by the stiff elastic feathers of the tail: the hoopoe, an inhabitant of southern Europe, is also a creeper, but it pursues small reptiles and insect on the ground.

The sylvias and thick-billed birds are by much the most characteristic of Europe; to them belong our finest songsters.. The sylvias have soft beaks, and feed on insects and worms; the nightingale, thrush, blackbird, wren, the beccafico, the smallest of European birds, the warblers, white-throat, and others, are of this family. Thick-billed birds live on seed, as the goldfinch and other finches, linnets, larks, buntings, and crossbeaks.

Four species of fly-catchers are peculiar to

Europe, and five species of shrikes. Ravens, crows, jays, and magpies, are everywhere; the Alpine crow and nutcracker are in central Europe only. Compared with America the starling family is poor, and the woodpecker race still more so, yet we have six species, some of which are very beautiful. There is only one cuckoo entirely European, the other two kinds only come accidentally, and all are birds of passage. There are four species of the pigeon tribe; the ring-dove frequents the larch forests, and is migratory; the stockdove also leaves us in October; the biset or rock pigeon, supposed to be the origin from which the infinite variety of our domestic pigeons has sprung, flies in flocks, and makes its flimsy nest on trees and rocks; it is also found in the Da-ouria part of the Altaï chain. Of gallinaceous birds there are many; the only native pheasant is in the south-western parts of the continent; and the capercaily, extinct in the British forests, inhabits many parts of Europe, in Scandinavia especially it is plentiful as far as the pine-tree grows, which is nearly to Cape North, and also in the Russian forests. The hazel grouse frequents the pine and aspen forests in central and northern Europe, where the black cock also is plentiful. Five species of grouse and six of partridges afford abundance of game; four of the latter are confined to the southern parts of the continent, and so are the sand and pen-tailed grouse, which form a separate family; the former inhabits the sterile plains of Andalusia and Granada, and the latter the stony uncultivated parts of France, southern Italy, and Sicily. The ortigis gibraltarica is a peculiar bird allied to the grouse family, found in the south of Europe only.

European waders are very numerous, and among them there are specimens of all the genera: woodcocks, snipes, plovers, curlews, and grebes, are very abundant, and herons of various species; three of them are egrets or crested herons, and the common heron now assembles on the tops of trees unmolested, since the progress of agriculture has rendered the country unfit for hawking. Several cranes and storks, and two species of ibis, are European: the flamingo is met with in the south-eastern parts of the continent, and in the maremme on the east coast of Italy. Many of the wading tribes, however, migrate in winter. The stork, so great a favourite in Holland that it is specially protected, is a wanderer; it retreats to Asia Minor, and on the return of summer resumes its old nest on a chimney-top, breeding in both countries. Europe is particularly rich in web-footed birds; there are four species of wild swans, four of wild geese, and more than 30 of the duck tribe, including the inhabitants of the Arctic seas.

BIRDS OF ASIA AND THE INDIAN ARCHIPELAGO.

European birds are widely spread over Asia; most of the Arctic sea-fowl frequent its northern coasts: between 50 and 60 European birds are also Siberian, and there are above 70 European species in Japan and Corea, which probably are also inhabitants of Siberia and the Altaï Mountains, and several are identical with the birds of North America; so that the same affinity prevails in the feathery tribes of the Arctic regions as in the vegetable productions.

Asia Minor is a country of transition, and many European birds inhabit the Caucasus, the shores of the Caspian Sea, and Persia. Moreover these warmer climates are the winter-quarters of various European species.

In Asia Minor, and especially in Armenia, the number and variety of birds is very great; large eagles, vultures, falcons, buzzards, quails, partridges, starlings, herons, storks, cranes, legions of Arctic grebes, swans, wild geese, ducks, and pelicans, are natives of these countries; besides singing-birds, the nightingale, the constant theme of the poet's song, abounds in Persia: hawks are trained for hunting deer in that country, and the Asiatic partridges, or francolins, more vividly coloured than ours, differ also in having beaks fitted for digging up bulbous roots, which is their food in the deserts.

Farther east the types become more Indian; the great peninsulas on each side of the Ganges are the habitations of the most peculiar and the most gorgeous of birds. Many species, and some entire genera, of kingfishers are here, of the gaudiest colouring; the plumage of the fly-catchers has the richest metallic lustre; and the shrikes, of a sober hue with us, are there decked in the brightest colours: the drango has a coat of ultramarine, and the calyptomene has one of emerald green.

The large-beaked climbing-birds are singularly handsome. The great green parrot, so easily taught to speak, has inhabited the Indian forests and the banks of the Ganges time out of mind, with a host of family connections and congeners of every colour; not one species of these, or indeed of the whole parrot tribe, is common to Asia, Africa, America, or Australia, nor even to any two of these great continents. They are vividly coloured in India, but the cuckoo tribe rivals them; several genera of these birds exist nowhere else, as the large-beaked malcahos, the coucals with their stiff feathers, and the couroucous or trogons, dressed in vermilion and gold; the last, however, also inhabit other tropical climates.

Eastern Asia is distinguished by the variety of its gallinaceous birds and the gorgeousness of their plumage. To this country we owe some of our domestic fowls; the cock and hen, and two species of peacock, are wild in the woods in India and Ceylon. The polyplectron, the only bird of its kind, and the trogopons, are Indian; and some of the most brilliant birds of the East are among the pheasant tribe, of which five species are peculiar to China and Tibet. There are various species of the horned pheasant in the Himalaya, and one whose feathers have a metallic lustre. The gold, the silver, and Reeves' pheasant, the tail-feathers of which are four feet long, belong to China. The lophophorus refulgens, and some others of that genus, are altogether Indian.

The pigeons also are very splendid in their plumage; they mostly belong to China and Japan; those in the Birman Empire are green.

It would be vain to enumerate the fine birds that range in the forests, or fish in the rivers of the Asiatic continent, yet the birds of the Indian Archipelago far surpass them in splendour of plumage; these islands indeed are the abode of the most gorgeously arrayed birds in existence. Even in Java and Sumatra, though most similar to India in their winged inhabitants, there are many peculiar, especially 12 or 13 species of the climbing tribe, and several of the honey-sucking kind; but the dissimilarity increases with the distance, as in New Guinea and its islands, where the honey-sucking genera are developed in novel forms and sumptuous plumage.

In the various islands of the archipelago there are altogether at least 15 genera, with their numerous species, found there only. There are the cassicans, which resemble jays, with plumage of metallic lustre; the only two species of pirolls, one bright violet, the other of brilliant green; various species of calaos with large horned beaks, oriols of vivid colours, the swallow that builds the edible nest, and every variety of birds of paradise; the most numerous and splendid sylvans, and all the species but one of the philedons or honey-sucking birds with tongues that end in a brush. The pigeons are peculiarly beautiful and numerous, but limited in their abode. The gouroa, or great crowned pigeon, the largest of its tribe, is an inhabitant of Borneo. Each island has its own species of louries, which exist nowhere else; many peculiar paroquets and cockatoos, couroucous or trogons, coucals, and the barbu, with huge beaks, are all peculiar to these islands. Even the partridges have thrown aside their grave colours and assumed the vivid hues of the tropics, as the green and tufted cryptonex. But the other gallinaceous birds far surpass them, as the fire and the argus pheasant, and two or three species still more splendid, with a host of other birds already known, and multitudes which Europeans never have seen, in the deep jungles and impenetrable forests of these unexplored islands. The casuary, a bird akin to the ostrich, without the power of flying, but fleet in its course, has a wide range in these countries, and, though destitute of beauty, is interesting from its peculiar location and the character of the whole race.

AFRICAN BIRDS.

A great number of European birds are also inhabitants of Africa, and many migrate there in winter, yet the birds of this continent are very peculiar and characteristic; those in the north and north-east, and at the Cape of Good Hope, are best known, but the greater part of tropical Africa is still unexplored. It may be observed, generally, that the tropical birds differ from those of North Africa, but are, with a few exceptions, the same with those in the southern part of the continent, and the whole of Africa south of the desert differs in species from those of north and western Africa and from Europe. Moreover, there is a strong analogy, though no affinity, between the birds of Africa and America in the same parallels of latitude: there is not a single perching bird common to the two, though some of the rapacious are in both.

There are 59 species of birds of prey, of which a few are also European. The secretary bird is the most singular of this order: it is a vulture which preys upon serpents at the Cape of Good Hope, in Abyssinia, and other parts of the continent. Africa possesses at least 300 species of the passerine order, of which 10 genera, with all their species, are peculiarly its own. The swallows are more beautiful than ours, especially the cecropis striata, with two tail-feathers twice as long as its body. Many kingfishers, the most beautifully coloured of their brilliant race, frequent the lakes and rivers: four species of hoopoes, one of which visits Europe in summer, are natives; and the honey-birds, the representatives of the humming-birds of South America, are peculiarly African. They abound at the Cape of Good Hope, where the nectaries of proteas and other plants produce the saccharine juice which is their food. The malurus Africanus, and many other singing-birds for the most part unknown elsewhere, inhabit the forests. The canary-bird is confined to the Canary Islands; its song differs in different parts, and even in two adjacent districts: there are, however, other instances of this. The capirote, also indigenous in the Canary Islands, is a finer songster, but it cannot be tamed. Various shrikes are peculiar to Africa, but the species known as the grand baratra is confined to the Azores. There are several birds of the raven tribe, or nearly akin to them: as the lampratornis superba; another with purple wings, the buphaga, the only bird of its genus; and several species of the calaos. The weaving-bird, or textor, is one of the most remarkable of the graminivorous tribe; it weaves its nest with grass and twigs very dexterously: one brought to Europe wove a quantity of thread among the wires of its cage, with great assiduity, into a strong texture. The widow-bird, the calious, the blue bee-eater, and all the fly-catching touracous, with many species of woodpeckers, are found nowhere else. The parrots and paroquets, which swarm in the tropical forests, from the size of a hen to that of a sparrow, are of original forms. The trogons, or couroucous, the most beautiful of the large-beaked climbing-birds, are the same as in Asia; but the barbu and the four species of barbicans are altogether African, and so are some of the cuckoos. Among the latter are two species of the cuculus indicator, so named from indicating where the bees have their nests; one is peculiar to Abyssinia, the other to the interior at the Cape of Good Hope: and mocking-birds are spread over a wide extent of this continent.

There are at least 13 species of African pigeons; and to Africa we are indebted for the guinea-fowl, of which there are three or four kinds: it wanders in flocks of hundreds among the brushwood on the banks of rivers and lakes in Numidia and all the tropical regions, and they are even more abundant in Madagascar. Many grouse and partridges are peculiar, especially the gangas, of which there are five species: some go in coveys, and others traverse the deserts in flocks of many hundreds. The sand-grouse, one of this family, is much more abundant on the arid deserts of North Africa than in Europe; and the partridges of this country are francolins which feed on bulbous roots.

The ostrich takes the wide range of Africa and Arabia; the bird of the desert, and bustards, also wanderers in the plains, are numerous: the most peculiar are the rhaad and the otis kori, in South Africa, five feet high, and remarkable for the brilliancy of its eye.

Waders of infinite variety inhabit the rivers, lakes, and marshes—woodcocks, snipes, plovers, storks, cranes, herons, and spoonbills. The most peculiar are the dromes and marabous, whose feathers form a considerable article of commerce; the cream-coloured plover, the scopus or umbrette, the water-treader of Abyssinia, and the tantalus tribe, among which is the falcinellus, known in Africa only, and the ibis, once held sacred in Egypt, and frequently found in mummies in the catacombs.

Swimming-birds are no less numerous: the bernicla cyanaptera is a goose peculiar to Shoa; the rhynchops and pelicans, several of the duck kind or birds allied to them, are found nowhere else.

BIRDS OF NORTH AMERICA.

Of 471 species of North American birds, about 100 are also found in Europe, the greater number of which are water-fowl, and those common to the northern coasts of both continents. The sea-fowl on the North Pacific and Behring's Straits are very much the same with those in the Greenland seas and the North Atlantic, but the great awk or penguin, with featherless wings, still exists on the North Pacific, and the great albatross, seldom seen in the North Atlantic, frequents Behring's Straits and the western coasts of North America in immense flocks. It is almost universal in the Pacific and in the stormy regions towards each pole. Like Mother-Cary's-chickens, it is a bird of the storm, sailing calmly on its huge wings in the most tremendous tempests, and following a ship a whole day without resting on the waves: it is the largest of sea-fowls; some measure 17 feet from tip to tip of the wings.

There is no vulture common to the two continents, but there are five eagles, half of the other birds of prey, a fourth part of the crow tribe, several waders and web-footed birds which inhabit both; yet the general character of North American birds is different from that of European: 81 American generic forms and two families are not found in Europe.

The humming-birds are altogether American; only four species are in North America; one of these is migratory, and another is common to South America. The parrot family, distributed with generic differences in almost all tropical countries, has but one representative here, which lives in the forests of the Carolinas. Europe has nothing analogous to these two families. It is singular that a country with so many rivers and lakes should possess only one kingfisher. The woods are filled with many species of creeping birds, and there are 68 peculiar species of sylvias and fly-catchers; among others the todus viridis, which forms a genus by itself. Ravens, crows, pies, and jays abound, and there are 13 species of starlings. The finch tribe are very numerous, and there are 16 species of woodpeckers, as might be expected in a country covered with forests. Of pigeons there are eight species, but individually they are innumerable, especially the columba migratoria, which passes over Canada and the northern States in myriads for successive days twice in the year. The poultry-yard is indebted to North America for the turkey, which there ranges wild in its native woods and attains great size. There are no partridges, and of 13 American species of grouse three are European, a family which exists in every country under different forms. The vast expanse of water and marshy ground makes North America the home of water-fowl and waders without end. Most of the waders and graminivorous birds are migratory; in winter they find no food north of the great lakes, where the ground is frozen upwards of six months in the year. Many pass the winter in California, as storks and cranes; wild geese cover acres of ground near the sea, and when they take wing their clang is heard far off. Blackbirds are as numerous; even gulls and other northern sea-fowl come to the coasts of California, and indeed to the shores of all the north and temperate Pacific.

It may be said generally that, with regard to the web-footed tribe, North America possesses specimens of all the genera of the old world and many peculiarly its own. The table-land of Mexico has some peculiar forms, and some species of swimming-birds found only in more northern latitudes; but, except the ampelidæ, there are representatives of every group of North and South America.

BIRDS OF SOUTH AMERICA.

The tenants of the air in South America differ more from those in North America than these do from the birds of Europe: there are not more than 50 or 60 species in common. South America has a greater variety of original forms than any other country; more than 25 genera with all their species inhabit that country only; of the passerine family alone there are at least 1000 species, all peculiar. The vultures are of different genera from those in Europe: the condor of the Andes is the largest of these; it is so fierce that it even attacks the puma, the lion of America; it frequents the highest pinnacles of the Andes in summer, and soars to the height of 15,000 feet above the earth. In winter it descends in groups to feed on the plains and sea-shore; and, like all the vulture race, it possesses the faculty of descrying a dead or dying animal long before it is itself visible in the air: it never goes beyond the isthmus of Panama: the condor of California is a smaller bird. The three species of the vultur papa, or king of the vultures, are remarkable for the bright blue and vermilion colour of the head and neck; the black vulture lives in large assemblies on the tops of high trees in the sylvas, and another numerous species prey on animals in the llanos. Many other rapacious birds are peculiar to this continent; the burrowing owl, so common in the Pampas and Chili, is one of these. The guachero forms a genus by itself; it is of the size of a common fowl, with the form and beak of a vulture, and is the only instance known of a nocturnal bird feeding on fruit. It is confined to a limited district of Cumana, and shuns the light: incredible numbers have taken possession of a dark cavern in the valley of Caripa, where they are killed in thousands every year by the Indians for their fat.

The troupials represent our oriols, the baratras and becardes our shrikes, while the tangaras partake of the form both of the shrike and pie, which last, with all the rest of the crow family, have various representatives in this country. Swallows, or birds allied to them, are numerous, and many that live on the honeyed juice of flowers, like the humming-bird, so peculiarly characteristic of South America: 150 species of humming-birds, from the size of a wren to that of a humble-bee, adorn the tropical regions of Brazil and Guiana. This family, so entirely American, has a range from the Straits of Magellan to the 38th parallel of N. lat., and even to Cook's Straits. There is only one South American humming-bird which is also permanent in the United States, and only two are found in Central America: many of them are migratory; they come in multitudes to North Chili in summer and disappear in winter. The climbing-birds, with large bills, are mostly confined to the tropical forests, which swarm with peculiar races of parrots, paroquets, and macaws, and with whole families of birds not to be seen elsewhere: as the vividly-coloured toucan, with its huge beak; the araucari, which lives on the fruit of the araucari pine; some peculiar species of the gorgeous trogons or couroucous; the tomalias, birds related to the cuckoo tribe; and the jacmars, which represent the woodpeckers.

The gallinaceous family is totally different from that in the Indian forests; the guan or penelope, related to the pheasant, and the tinamous, something of the grouse kind, supply their place, together with various alectors, which run after lizards and snakes on the plains, or feed on insects on the banks of rivers. Some of them have a horny substance on the wings for striking their prey; the most peculiar of these alectors are the agami or trumpet bird, the kamichi, and the caziama, of Brazil. No country can be compared with South America for the number of original forms of birds, far beyond even being mentioned in a book not entirely devoted to natural history.

The ostrich with three toes, or struthia rhea, ranges, like all its congeners, over a wide extent of country. It is everywhere from the Silvas to the Rio Negro, which bounds the Pampas of Buenos Ayres; while the struthia Darwinii has the plains of Patagonia to the Straits of Magellan for its residence.

The water-fowl and waders in this land of rivers are beyond number; millions of flamingoes, spatules, cormorants, herons, fishing-falcons, and rhynchops, follow the fish that go up the rivers to spawn; nor are gulls wanting where fish are to be found: a little snow-white heron walks on the back and over the head of the crocodile while it sleeps. The water-fowl are almost all peculiar; the few that are excepted are North American. Eight or nine genera belonging to the warm climates of the old world, are here under new forms, and the number of specific forms of the same genus is greater than in any other country. The tantalus ruber inhabits Cayenne; the ardea helias and scalopax are the most peculiar of the herons.

Ducks migrate in immense flocks, alternately between the Orinoco and the Amazons, on account of the greater supply of fish afforded by the floods of these rivers, which take place at intervals of six months from each other. Between the tropics the vicissitudes of drought and humidity have much influence on the migration of birds, because the supply of their food depends upon these changes.

If any thing more were required to show the partial location of birds, the Galapagos Archipelago might be mentioned: of 26 specimens shot by Mr. Darwin, 25 were peculiar, though bearing a strong resemblance to American types; some birds were even confined to particular islands; and the gulls, one of the most widely dispersed families, are peculiar. But on this comparatively recent volcanic group, only 500 miles distant from the coast of America, everything is peculiar, birds, plants, reptiles, and fish, and though under the equator, all have sober covering.

The coasts of Peru and northern Chili are not rich in birds, but in southern Chili there are many humming-birds, parrots, giant storks, flamingoes, peculiar ducks and geese: and there begins that inconceivable quantity of sea-fowl that swarm on the seas and coasts of the Antarctic regions. The black rayador, or rhynchops nigra, has been seen in a dense mass seven miles long; shags fly in an unbroken line two miles long. Pelicans, sea-ravens, gulls, petrels, and many others cover the low islands and coasts of the mainland, and those of Tierra del Fuego.

In the Antarctic seas petrels take place of our gulls; seven species of them inhabit these high southern latitudes in prodigious numbers. A flock of what was supposed to be the young of the kind known as the Cape pigeon, was estimated to have been from six to ten miles long, and two or three miles broad, which absolutely darkened the air during the two or three hours they were flying over the discovery ships. The white petrel, a most elegant bird, never leaves the ice, and consequently is never seen north of the Antarctic circle in summer. Three species of penguin inhabit these seas; the largest, which is a rare and, for the most part, solitary bird, lives on the pack-ice, and weighs from 60 to 70 pounds. The other two species are smaller and gregarious; they crowd the snow-clad islands in the high southern latitudes in myriads: every ledge of rock swarms with them, and on the shore of Possession Island, close to Victoria Land, it was difficult to pass through the multitudes. They are fine, bold birds, pecking and snapping with their sharp bills at those who venture among them. They can scarcely walk, and cannot fly, but they skim along the snow, and swim rapidly, even under water, and the noise they make baffles all description. Two species of albatross breed in the Antarctic Islands; a kind of skua gúll, which robs their nests; and a goose which, like the eider-duck, makes its nest with the down from its breast. Few land-birds are met with within the Antarctic circle: there are but seven or eight species in the Auckland Islands, mostly New Zealand birds; among others, the tooa or tui, and an olive-coloured creeper, the choristers of the woods. One only was found in Campbell Island.

Many generic forms are the same at the two extremities of the globe, yet with distinct specific differences. Sea-fowls are more excursive than other birds, but even they confine themselves within definite limits, so that the coasts may be known from their winged inhabitants.

AUSTRALIAN BIRDS.

The Australian birds are in many respects as singular as the quadrupeds and plants of that country: a white falcon is among its birds of prey, a black swan among its water-fowl, and of 45 genera, 35 are purely Australian. The passeres are so original, that many new genera have been found. The cassican, a handsome bird of bright colours, approaching somewhat to the crow family, the choucalcyon, the golden and black oriole, and one species of phelidon, are peculiar Australian. The menura superba, or lyre-bird, from the resemblance its outspread tail bears to the form of the ancient lyre, is the only bird of its genus, and the only one which approaches the character of the gallinaceous family, of which none have been discovered in the Australian continent. Here are many specific kinds of cuckoos, as the coucals and the scythrops, the only bird of its genus. Woodpeckers there are none. The parrots, paroquets, and cockatoos, which live in numerous societies, are all peculiar, especially the black cockatoo, which is found here only; it is not so gregarious, but even more suspicious than the white cockatoos, which have a sentinel to warn them of danger. Chious, with huge bills like the toucan, satin-birds, pigeons and doves of original forms, abound; and the cereops goose is no less peculiar among the web-footed tribe. The desert plains of this great continent are allotted to the emu, a large struthia, like its congener the ostrich, incapable of flight, and once very plentiful, but now in progress of being extirpated or driven by the colonists to the unexplored regions of the interior.

The apteryx, a bird of the same family, still lingers in New Zealand, but it is on the verge of extinction, and probably owes its existence to its nocturnal and burrowing habits. It is one of those anomalous creatures that partakes of the character of several others; its head is in some degree like that of the ibis, with a long slender bill, fitted for digging into the ground for worms and grubs; its legs and feet resemble those of the common fowl, with a fourth toe or spur behind, in which it differs from its congeners; and its wings, if wings they can be called, are exceedingly small. In a specimen, whose body measured 19 inches, the wings, stripped of the feathers, were only an inch and a half long, ending in a hard horny claw three inches long. The comparatively small wings are characteristic of the whole family: the rhea and ostrich have the largest, which though unavailing in flight, materially aid their progress in running; the wings of the emu and apteryx serve only as weapons of defence: the whole tribe also defend themselves by kicking. No animals have a more remarkable geographical distribution than this family, or show more distinctly the decided limits within which animals have originally been placed. These huge birds can neither fly nor swim, consequently they could not have passed through the air or the ocean to distant continents and islands. There are five distinct genera, to each of which very extensive and widely separated countries have been allotted: the ostrich is spread over Africa, from the Cape of Good Hope to the deserts of Arabia; two species of the rhea range over the plains of the Pampas and Patagonia, in South America; the continent of Australia is the abode of the emu; the cassowary roves over some of the large islands of the Indian Archipelago; and the apteryx dwells in New Zealand. The dodo, a very large bird of the struthia kind, extirpated by the Dutch navigators, once inhabited Mauritius and the adjacent island of Don Rodriguez. The deinornis giganteus, a bird 10 feet high, has been recently extinguished in New Zealand, if there be not still some lingering in the unexplored part of that wide country, the only one that has contained two genera of this family of birds. Bones, not fossilized, but in the natural state, have been found of six species of this extraordinary bird, and brought to England; and a complete skeleton of the deinornis giganteus has been arranged by Professor Owen, the distinguished comparative anatomist, to whom we are indebted for a very interesting account of it. A small portion of a large bone was examined by him, and the result was one of those triumphs of science which characterize genius; he boldly pronounced it to be the bone of a bird—of the ostrich kind, and his decision has since been abundantly confirmed by the subsequent discovery of the bones and part of the egg of the bird.

The struthia family live on vegetables; the form of those that had their home in New Zealand shows that they had fed on the edible roots of the fern which covers that country; and as no quadruped excepting a rat is indigenous in New Zealand, though 700 miles long, and in many places 90 wide, these birds could have had no enemy but man, the most formidable of all.

The beautiful and sprightly tui, or parson bird, native in New Zealand, is jet black, with a white tuft on its breast, and so imitative that it can be taught to repeat whole sentences. There are parrots and paroquets, vast numbers of pigeons, fine warblers, many small birds, and a great variety of water-fowl, amongst others a cormorant, which, though web-footed, perches on the trees that overhang the streams and sea, watching for fish; and a snow-white frigate-bird, that pounces on them from a great height in the air. Altogether there are at least 84 species of birds that inhabit these islands.

CHAPTER XXX.

DISTRIBUTION OF MAMMALIA THROUGHOUT THE EARTH.

CARBONIC acid, water, and ammonia, contain the elements necessary for the support of animals, as well as of vegetables. They are supplied to the graminivora in the vegetable food which is converted into animal substance by their vital functions.

Vitality in animals, as in vegetables, is the power they have of assimilating their food, a process independent of volition, since it is carried on during sleep, and is the cause of force. Animals inhale oxygen with the air they breathe; part of the oxygen combines with the carbon contained in the food, and is exhaled in the form of carbonic acid gas. With every effort, with every breath, and with every motion, voluntary or involuntary, at every instant of life, a part of the muscular substance becomes dead, separates from the living part, combines with the remaining portion of inhaled oxygen, and is removed. Food, therefore, is necessary to compensate for the waste, to supply nourishment, and to restore strength to the nerves, on which all vital motion depends; for by the nerves volition acts on living matter. Food would not be sufficient to make up for this waste, and consequent loss of strength, without sleep; during which voluntary motion ceases, and the undisturbed assimilation of the food suffices to restore strength, and to make up for the involuntary motion of breathing, which is also a source of waste.

The perpetual combination of the oxygen of the atmosphere with the carbon of the food, and with the effete substance of the body, is a real combustion, and is supposed to be the cause of animal heat, because heat is constantly given out by the combination of carbon and oxygen; and, without a constant supply of food, the oxygen would soon consume the whole animal, except the bones.

Graminivorous animals inhale oxygen in breathing, they also take it in by the pores of the skin; and as vegetable food does not contain so much carbon as animal food, they require a greater supply to compensate for the wasting influence of the oxygen; therefore, cattle are constantly eating. But the nutritious parts of vegetables are identical in composition with the chief constituents of the blood; and from blood every part of the animal body, and even a portion of the bones, is formed.

Carnivorous animals have not pores in the skin, therefore their supply of oxygen is from their breath only; and, as animal food contains a greater quantity of carbon, they do not require to eat so often as animals that feed on vegetables. The restlessness of carnivorous animals, when confined in a cage, is owing to the superabundance of carbon in their food. They move about continually to quicken respiration, and by that means procure a supply of oxygen to carry off the redundant carbon.

The quantity of animal heat is in proportion to the amount of the oxygen inspired in equal times. The heat of birds is greater than that of quadrupeds, and in both it is higher than the temperature of amphibious animals, and fishes, which have the coldest blood. On these subjects we are indebted to Professor Liebig, who has thrown so much light on the important sciences of animal and vegetable chemistry.

The mammalia consist of nine orders of animals, which differ in appearance and in their nature; but they agree in the one attribute of suckling their young. These orders are—the quadrumana, animals with four hands, as monkeys and apes; cheiroptera, animals with winged hands, as bats; carnivora, that live on animal food, as the lion and tiger; rodentia, or gnawers, as beavers, squirrels, mice; edentata, or toothless animals, as ant-eaters and armadilloes; pachydermata, or thick-skinned animals, as the elephant, the horse; ruminantia, animals that chew the cud, as cows, sheep, deer; cetaceæ, as whales, dolphins, and phocæ.

The distribution of animals is guided by laws analogous to those which regulate the distribution of plants, insects, fishes, and birds. Each continent, and even different parts of the same continent, are centres of zoological families, which have always existed there, and nowhere else; each group being almost always specifically different from all others.

Food, security, and temperature have no influence, as primary causes, in the distribution of animals. The plains of America are not less fit for rearing oxen than the meadows of Europe; yet the common ox was not found in that continent at the time of its discovery; and, with regard to temperature, this animal thrives on the llanos of Venezuela and the pampas of Brazil as well as on the steppes in Europe. The horse is another example: originally a native of the deserts of Tartary, he now roams wild in herds of hundreds of thousands on the grassy plains of America, though unknown in that continent at the time of the Spanish invasion. The stations which the different families now occupy must have been allotted to them as each part of the land rose above the ocean; and because they have found in these stations all that was necessary for their existence, many have never wandered from them, notwithstanding their powers of locomotion; while others have migrated, but only within certain bounds.

The Arctic regions form a district common to Europe, Asia, and America. On this account, the animals inhabiting the northern parts of these continents are sometimes identical, often very similar; in fact, there is no genus of quadrupeds in the Arctic regions that is not found in the three continents, though there are only 27 species common to all, and these are mostly fur-bearing animals. In the temperate zone of Europe and Asia, which forms an uninterrupted region, identity of species is occasionally met with; but for the most part marked by such varieties in size and colour as might be expected to arise from difference of food and climate. The same genera are sometimes found in the intertropical parts of Asia, Africa, and America, but the same species never; much less in the south temperate zones of these continents, where all the animals are different, whether birds, beasts, insects, or reptiles; but in similar climates analogous tribes replace one another.

Europe has no family and no order peculiarly its own, and many of its species are common to other countries; consequently the great zoological districts, where the subject is viewed on a broad scale, are Asia, Africa, Oceanica, America, and Australia; but in each of these there are smaller districts, to which particular genera and families are confined. Yet when the regions are not separated by lofty mountain-chains, acting as barriers, the races are in most cases blended together on the confines between the two districts, so that there is not a sudden change.

EUROPEAN ANIMALS.

The character of the animals of temperate Europe has been more changed by the progress of civilization than that of any other quarter of the globe. Many of its original inhabitants have been extirpated, and new races introduced; but it seems always to have had various animals capable of being domesticated. The wild cattle in the parks of the Duke of Hamilton and the Earl of Tankerville are the only remnants of the ancient inhabitants of the British forests, though they were spread over Europe, and perhaps were the parent stock from which the European cattle of the present time have descended; though the bison, or euroch, a race nearly extinct, and found only in the forests of Lithuania and the Caucasus, may have some claim to the pedigree. Both races are supposed to have come from Asia. The musmon, which exists in Corsica and Sardinia, is said to be the origin from which our sheep sprung. The pig, the goat, the fallow-deer, and red-deer, have been reclaimed, and also the reindeer, which cannot strictly be called European, since it also inhabits the northern regions of Asia and America. The cat is European; and altogether eight or ten species of tamed quadrupeds have sprung from native animals.

There are still about 180 wild land-animals in Europe: 45 of these are also found in western Asia, and nine in northern Africa. The most remarkable are the reindeer, elk, red and fallow deer, the roebuck, glutton, lynx, polecat, several wild-cats, the common and black squirrels, the fox, wild boar, wolf, the black and the brown bear, eight species of weazels, and seven of mice. The otter is common; but the beaver is now found only on the Rhine, the Rhone, the Danube, and some other large rivers; rabbits and hares are numerous; the hedgehog is everywhere; the porcupine in southern Europe only; the chamois, yzard, and ibex in the Alps and Pyrenees. Many species of these animals are widely distributed over Europe, generally with variations in size and colour. The chamois of the Alps and Pyrenees, though the same in species, is slightly varied in appearance; and the fox of the most northern parts of Europe is larger than that in Italy, with a richer fur, and somewhat different colour.

Some European animals are much circumscribed in their locality. The ichneumon is peculiar to Spain; a peculiar species of stag and the musmon are confined to Corsica and Sardinia; there are a weazel and bat which inhabit Sardinia only; and Sicily has several peculiar species of bats and mice. There is only one species of monkey in Europe, which lives on the rock of Gibraltar, and is supposed to have been brought from Africa. All the indigenous British quadrupeds now existing, together with the hyæna, tiger, bear, and wolf, whose bones have been found in caverns, came from Germany before England was cut off from the continent by the British Channel; but the greater number have perished. Ireland was separated by the Irish Channel before all the animals had migrated across England; so that our squirrel, mole, polecat, dormouse, and many smaller quadrupeds, never reached the sister island.

ASIATIC ANIMALS.

Asia has a greater number and a greater variety of wild animals than any country, except America, and also a larger proportion of those that are domesticated. Though civilized from the earliest ages, the destruction of the animal creation has not been so great as in Europe, owing to the inaccessible height of the mountains, the extent of the plains and deserts, and, not least, to the impenetrable forests and jungles, which afford them a safe retreat: 288 mammalia are Asiatic, of which 186 are common to it and other countries; these, however, chiefly belong to the temperate zone.

Asia Minor is a district of transition from the fauna of Europe to that of Asia. There the chamois, the bouquetin, the brown bear, the wolf, fox, hare, and others, are mingled with the hyæna, the angora goat, which bears a valuable fleece, the argali or wild sheep, the white squirrel, peculiar deer; and even the Bengal royal tiger is sometimes on Mount Ararat, and is not uncommon in Azerbijan and the mountains in Persia.

Arabia is inhabited by the hyæna, panther, jackal, wolf, and musk-deer. Antelopes and monkeys are found in Yemen and Aden. Most of these are also indigenous in Persia. The wild ass, a handsome spirited animal of great speed, and so shy that it is scarcely possible to come near it, wanders in herds over the deserts in both countries. It is also indigenous in the Indian desert, and especially in the Run of Cutch: "the wilderness and the barren lands are his dwelling."

The table-lands and mountains which divide eastern Asia almost into polar and tropical zones, produce as great a distinction in the character of its indigenous fauna. The severity of the climate in

Siberia renders the skins of its numerous fur-bearing animals more valuable. These are reindeer, elks, wolves, the large white bear, that lives among the ice on the Arctic shores, several other bears, the lynx, various kinds of martens and cats, the common, the blue, and the black fox, the ermine, and sable. The fur of these last is much esteemed, and is inferior only to that of the sea-otter, which inhabits the shores on both sides of the Northern Pacific.

With the exception of the jerboa, which burrows in sandy deserts, on the table-land and elsewhere, all the Asiatic species of gnawers are confined to Siberia. The most remarkable of these is the flying squirrel. The Altai Mountains teem with wild animals, besides many of those mentioned. There are large stags, sloths, some peculiar weazels, the argali, and the musmon, or wild sheep, the same with that in Sardinia. The wild goat of the Alps is found in the Sayansk part of the chain; the glutton and musk-goat in the Baikal; and in Da-Ouria the red-deer and a peculiar antelope. The Bengal tiger and the felis-irbis, a species of panther, wander from the Celestial Mountains to the Altai chain and southern Siberia; and the tiger is met with even on the banks of the Obi, and also in China, though in the northern regions it differs considerably from the same species in Bengal. The tapir, and many of the animals of the Indian Archipelago, are in the southern provinces of the Chinese empire; but its fauna is little known. It is, however, probable that in the northern parts it resembles that of the Altai Mountains and Siberia. The animals of Japan have a strong analogy to those of Europe: many are identical, or slightly varied, as the badger, otter, mole, common fox, marten, and squirrel. On the other hand, a large species of bear in the island of Jezo is analogous to the grizzly bear in the rocky mountains of North America. A chamois in other parts of Japan is similar to the chamois montana of the same mountains; and other animals native in Japan are the same with those in Sumatra; so that its fauna is connected with that of very distant regions.

A few animals are peculiar to the high cold plains of the table-land of eastern Asia: the dzigguetai, a very fleet animal, resembling both the horse and the ass, is peculiar to these Tartarian steppes; two species of antelopes inhabit the plains of Tibet, congregating in immense herds, with sentinels so vigilant that it is scarcely possible to approach them. The dzeran, or yellow goat, which is both swift and shy, and the handsome Tartar ox, are native in these wilds; also the shawl-wool goat and the manul, from which the Angora cat, so much admired in Persia and Europe, is descended.

The ruminating animals of Asia are more numerous and more excellent than those of any other part of the world; 64 species are native, and 46 of these exist there only. There are several species of wild oxen; one in the Burmese empire, and on the mountains of north-eastern India, with spiral twisted horns. The buffalo is native in China, India, Borneo, and the Sunda Islands; it is a large animal, formidable in a wild state, but domesticated universally in the East. It was introduced into Italy in the sixth century, and large herds now graze in the low marshy plains near the sea.

Various kinds of oxen have been domesticated in India time immemorial: the handsome Indian ox, with a hump on the shoulder, has been venerated by the Bramins for ages; the beautiful white silky tail of the domesticated Tartar ox, used in the East to drive away flies, was adopted as the Turkish standard; and the common Indian ox differs from all others in having great speed. Some other species of cattle have been tamed, and some are still wild in India, Java, and other Asiatic Islands. The Cashmere goat, which bears the shawl wool, is the most valuable of the endless varieties of goats and sheep of Asia; it is kept in large herds on the central table-land, on the northern declivities of the Himalaya, and in the upper regions of Bhotan, where the cold climate is congenial to it.

Twelve species of antelope and 20 of deer are peculiar to Asia, of which the musk-deer of the Himalaya is one; two species of antelopes have been mentioned as peculiar to the table-land, others are distributed in the islands.

Asia possesses eight native species of thick-skinned animals, including the elephant, horse, ass, camel, and dromedary, which have been domesticated from the time of the earliest scriptural records. The horse and camel are supposed to have existed wild in the plains of Central Asia, and the dromedary in

Arabia; though now they are only known as domestic animals. The Arabian and Persian horses have acknowledged excellence and beauty, and from these our best European horses are descended; the African horse, which was taken to Spain by the Moors, is probably of the same race.

The elephant has long been a domestic animal in Asia, though it still roams wild in formidable herds through the forests and jungles at the foot of the Himalaya, in other parts of India, the Indo-Chinese peninsula, and the islands of Sumatra and Ceylon, where it seems to be of a different species of those that are tame; the hunting elephant is esteemed the most noble. A rhinoceros with one horn is native on the continent.

There are 60 genera of Asiatic carnivorous animals, of which the royal tiger is the handsomest and the most formidable, its favourite habitation is in the jungles of Hindostan, though it wanders nearly to the limit of perpetual snow in the Himalaya, to the Persian and Armenian mountains, to Siberia and China. Leopards and panthers are common, and there is a maneless lion in Guzerat: the chitta, used in hunting, is the only one of the tigers capable of being tamed. The hyæna is found everywhere, excepting the Birman empire, in which there are neither wolves, hyænas, foxes, nor jackals. There are four species of carnivorous bears in India; that of Nepaul has valuable fur: the wild boar, hog, and dogs of endless variety, abound.

Toothless animals have only two representatives

in India; which, however, differ from all others except the African, in being covered with imbricated scales, which they can erect at pleasure.

The Indian Archipelago and the Indo-Chinese peninsula form a zoological province of a very peculiar nature, being allied to the faunas of India, Australia, and South America, yet having animals exclusively its own. Some groups of the islands have several animals in common, either identical or with slight variations, that are altogether wanting in other islands, which, in their turn, have creatures of their own. Many species are common to the Archipelago and the neighbouring parts of the continent, or even to China, Bengal, Hindostan, and Ceylon. Flying quadrupeds are a distinguishing feature of this archipelago, though they do not absolutely fly, but, by an extension of the skin of their sides to their legs, they take long leaps. Nocturnal flying squirrels, of several species, are common to the Malayan peninsula and the Sunda Islands, especially Java; and three species of flying lemurs inhabit Sunda, Malacca, and the Pelew Islands. Besides these, there are the frugivorous bats, which really fly, and differ from bats in other countries in living upon vegetable food.

A hundred and eighty species of the ape and monkey tribe are entirely Asiatic: monkeys are found only on the coast of India, Cochin-China, and the Sunda Islands; the long-armed apes or gibbons are in the Sunda Islands and the Malayan peninsula; and the pongos or orang-outang are natives of Sumatra and Borneo. The simayang, a very large ape of Sumatra and Bencoolen, goes in large troops, following a leader, and makes a howling noise at sunrise and sun-set that is heard miles off. Sumatra and Borneo are the peculiar abode of the orang-outang, which in the Malay language means the "man of woods," and of all its kind, except perhaps the chimpanzee of Africa and the kahau of the Malayan peninsula, approaches nearest to man. It has never spread over the islands it inhabits, though there seems to be nothing to prevent it, but it finds all that is necessary within a limited district. The orang-outang and the long-armed apes have extraordinary muscular strength, and swing from tree to tree by their arms.

The Malays have given the name of orang or man to the whole tribe, on account of their intelligence as well as their form.

A two-horned rhinoceros is peculiar to Java, of a different species from the African, also the felis macrocelis, and a very large bear; there are only two species of squirrels in Java, which is remarkable, as the Sunda Islands are rich in them. The royal tiger of India and the elephant are found only in Sumatra, and the babi-roussa or hog-deer lives in Borneo; but these two islands have many quadrupeds in common, as a leopard, the one-horned rhinoceros, the black antelope, some graceful miniature creatures of the deer kind, the tapir bicolor, also found in Malacca and India, besides a wild boar, an inhabitant of all the marshy forests from Borneo to

New Guinea. In the larger islands deer abound, from the size of a rabbit to that of the elk.

The anoa, a ruminating animal about the size of a sheep, and in appearance something between the buffalo and antelope, shy and fierce, goes in herds in the mountains of Celebes, where many forms of animals strangers to the Sunda Islands begin to appear, as some sorts of phalangers, or pouched quadrupeds. These new forms become more numerous in the Moluccas, which are inhabited by flying phalangers and other pouched animals, with scaly tails. In New Guinea there are kangaroos, the spotted phalanger, the pelandoe, the New Guinea hog, and the Papua dog, said to be the origin of all the native dogs in Australia and Oceanica, wild or tame.

The fauna of the Philippine Islands is analogous to that in the Sunda Islands. They have several quadrupeds in common with India and Ceylon, but there are others which probably are not found in these localities.

AFRICAN QUADRUPEDS.

The opposite extremes of aridity and moisture in the African continent have had great influence in the nature and distribution of its animals; and since by far the greater part consists of plains utterly barren or covered by temporary verdure, and watered by inconstant streams that flow only a few months in the year, fleet animals, fitted to live on arid plains, are far more abundant than those that require rich vegetation and much water. The latter are chiefly confined to the intertropical coasts, and especially to the large jungles and deep forests at the northern declivity of the table-land, where several genera and many species exist that are not found elsewhere. Africa has a fauna in many respects insulated from that of every other part of the globe; for although about 100 of its quadrupeds are common to other countries, there are 250 species its own. Several of these animals, especially the larger kinds, are distributed over the whole table-land from the Cape of Good Hope to the highlands of Abyssinia and Senegambia without the smallest variety, and many are slightly modified in colour and size. Ruminating animals are very numerous, though few have been domesticated: of these the ox of Abyssinia and Bornou is remarkable from the extraordinary size of its horns, which are sometimes two feet in circumference at the root; and the Galla ox of Abyssinia has horns four feet long. There are many African species of buffaloes; that at the Cape of Good Hope is a large, fierce animal, wandering in herds in every part of the country, even to Abyssinia: the flesh of the whole race is tainted with the odour of musk. The African sheep and goats, of which there are many varieties, differ from those of other countries; the wool of all is coarse, except that of the Merino sheep, said to have been introduced into Spain by the Moors from Morocco.

No country has produced a ruminating animal similar, or even analogous, to the giraffe, or camelopard, which ranges widely over South Africa from the northern banks of the Gareep, or Orange River, to the Great Desert. It is a gentle, timid animal, which has been seen in troops of 100. The earliest record we have of it is, that it graced the triumph of a Roman emperor.

Africa may truly be said to be the land of the antelope, which is found in every part of it, though chiefly on the table-land. Different species have their peculiar localities, while others are widely dispersed, sometimes with and sometimes without any sensible variety of size or colour. The greater number are inhabitants of the plains, while a few affect the forests. Sixty species have been described, of which at least 26 are found at the Cape of Good Hope and in the adjacent countries. They are of every size, from the pigmy antelope not larger than a hare, to the eland, which is larger than a calf. Timidity is the universal character of the race. Many are gregarious ; and the number in a herd is far too great even to guess at. Like all animals that feed in groups, they have sentinels ; and they are the easy prey of so many carnivorous animals, that their safety requires the precaution. At the head of their enemies is the lion, who lurks among the tall reeds at the fountain, to seize them when they come to drink. They are graceful in their motions, especially the spring-buck, which goes in a compact troop ; and in their march there is constantly some one which gathers its slender limbs together and bounds into the air.

Africa has only two species of deer, both belonging to the Atlas: one is the common fallow-deer of Europe.

The 38 species of rodentia, or gnawing quadrupeds, of this continent, live on the plains; and the greater part of them are leaping animals, as the gerboa capensis. Squirrels are rare, and all terrestrial.

There are five species of the horse kind in South Africa; of these the gaily-striped zebra, and the more sober-coloured quagga, of several species, wander in troops over the plains, often in company with ostriches. An alliance between creatures differing in nature and habits is not easily accounted for. The two-horned rhinoceros of Africa is different from that of Asia: there are certainly three, and probably five, species of these huge animals peculiar to the table-land. Dr. Smith saw 150 in one day near the 24th parallel of south latitude. The hippopotamus is exclusively African: multitudes inhabit the lakes and rivers in the intertropical and southern parts of the continent, and never change their abode. Elephants, differing in species from those in Asia, are so numerous, that 200 have been seen in a herd near Lake Chad. They are not now domesticated in Africa, and are hunted by the natives for their tusks. A wild hog and the hyrax are among the thick-skinned quadrupeds of this country. The monkey tribe is found in all the hot parts of Africa: peculiar genera are allotted to particular districts. The family of guenons is found in no part of the world but the Cape of Good Hope, the coasts of Loango and Guinea; the mandrills are peculiar to Guinea; and of the cynocephalus, or blue-headed ape, one species inhabits Guinea, others the southern part of the table-land, and one is met with everywhere from Sennaar to Cafraria. A very remarkable long-eared kind is found in Abyssinia; the margot is in North Africa, and the chimpanzee inhabits the forests of South Africa from Cape Negro to the Gambia. Living in society like all apes and monkeys, which are eminently sociable, it is easily tamed, and very intelligent. Baron Humboldt observes that all apes resembling man have an expression of sadness; that their gaiety diminishes as their intelligence increases.

Africa possesses the cat tribe in great variety and beauty; lions, leopards, and panthers are numerous throughout the continent; servals and viverrine cats are in the torrid districts; and the lion of the Atlas is said to be the most formidable of all. In no country are foxes so abundant. Various species inhabit Nubia, Abyssinia, and the Cape of Good Hope. The corsac is peculiar to the Cape. The long-eared fox, the famel of Kordofan, and some others, are found in Africa only. There are also various species of dogs, the hyæna, and the jackal.

Two species of toothless animals are African—the long-tailed manis, and the aard-vark, or earth-hog; both are covered with scales: they burrow in the ground, and feed on ants. Great flocks of a large migratory vampire-bat frequent the slave-coast. Altogether there are 26 species of African bats.

Multitudes of antelopes of various species, lions, leopards, panthers, hyænas, jackals, and some other carnivora, live in the oases of the great northern deserts; gerboas and endless species of leaping gnawers, rats and mice, burrow in the ground. The dryness of the climate and soil keeps the coats of the animals clean and glossy; and it has been observed that tawny and grey tints are the prevailing colours in the fauna of the North African deserts, not only in the birds and beasts, but in reptiles and insects. In consequence of the continuous desert soil from North Africa through Arabia to Persia and India, many analogous species of animals exist in those countries; in some instances they are the same, or varieties of the same species, as antelopes, leopards, panthers, jackals, and hyænas.

The fauna on the eastern side of the great island of Madagascar is analogous to that of India; on the western side it resembles that of Africa, though, as far as it is known, it seems to be a distinct centre of animal life. It has no ruminating animals; and the monkey tribe is represented by the lemures, which are characteristic of this fauna. A frugivorous bat, the size of a common fowl, forms an article of food.

AMERICAN QUADRUPEDS.

No species of animal has yet been extirpated in America, which is the richest zoological province, possessing 537 species of mammalia, of which 480 are its own, yet no country has contributed so little to the stock of domestic animals. With the exception of the llama (alpaca guanaco) and vicugna, the turkey, and perhaps some sheep and dogs, America has furnished no animal or bird serviceable to man, while it has received from Europe all its domestic animals and its civilized inhabitants.

Arctic America possesses almost all the valuable fur-bearing animals that are in Siberia; and they were very plentiful till the unsparing destruction of them has driven those yet remaining to the high latitudes, where the hunters that follow them are exposed to great hardships. Nearly 6,000,000 of skins were brought to England in one year. Of the large animals, the shaggy bison, the musk-ox, and the wapiti are peculiar. The musk-ox travels north to Parry's Islands; yet it never has been seen in Greenland or on the north-west coast of America. The range of the elk ends where the aspen and willow cease to grow. The rein-deer, living on lichens and mosses, wanders to the shores of the Polar Ocean: its southern limit in Europe is the Baltic, and in America the latitude of Quebec. The white bear, the largest and most formidable of his kind, inhabits the ice itself. The shaggy bison goes south to the Arkansas, and roams in herds of thousands over the prairies of the Mississippi, and on both sides of the Rocky Mountains. A marten called the prairie-dog is universal.

There are at least eight species of American dogs, several of which are natives of the far north. The logapus, or isatis, native in Spitzbergen and Greenland, is found in all the Arctic regions of America and Asia, and in some of the Kurile Islands. Dogs are employed to draw sledges in Newfoundland and Canada; and the Esquimaux travel drawn by dogs as well as by rein-deer. The dogs are strong and docile. The Esquimaux dogs were mute, till they learned to bark from dogs in our discovery ships.

There are 13 species of the ruminating genus in North America, including the bison, the musk-ox of the Arctic regions, the big-horned sheep, and the goat of the Rocky Mountains; but of the thick-skinned tribe, so useful to man, there are only some tapirs, and a creature allied to the hog. The horse, now roaming wild in innumerable herds over the plains of South America, was unknown there till the Spanish conquest. Some of the fur-bearing animals of the north never pass 65° N. lat., and the rest live in the pine-forests of Canada. The quadrupeds of the temperate zone are also distributed in distinct groups: those of the state of New York, consisting of about 40 species, are different from those of the Arctic regions, and also from those of South Carolina and Georgia; while in Texas another assemblage of species prevails. Numerous species of gnawers are scattered over the northern continent, especially squirrels; the grey squirrel is in thousands; but the racoon, the coatimondi, and the kinkajou are all natives of the southern States. The opossum, a pouched animal of an order peculiarly Australian, is found in Virginia, and everywhere between the great Canadian lakes and Paraguay; and two other animals of that order live in Mexico. There is a porcupine in the United States and Canadian forests which climbs trees. The bats are different from those in Europe, and, excepting two, are very local. The grizzly bear of the Rocky Mountains is the largest and most ferocious of American bears. The prong-buck antelope is everywhere in the western parts of the continent, from 53° N. lat. to Mexico and California; it is swifter than the fleetest horse, and migrates to the south in winter. In California there are ounces, polecats, the fallow-deer, the berenda (an animal peculiar to that country), and a deer of remarkable size and speed.

The high land of Mexico forms a very decided line of division between the fauna of North and that of South America; yet some North American animals are seen beyond it, particularly two of the bears, and one of the otters, which inhabits the continent from the icy ocean to beyond Brazil. On the other hand, the puma, jaguar, opossum, kinkajou, and peccari have crossed the barrier from South America to California and the United States.

In the varied and extensive regions of South America there are several centres of a peculiar fauna, according as the country is mountainous or level, covered with forest or grass, fertile or desert, but the mammalia are inferior in organization and size to those of the old world. The largest, most powerful, and perfect animals of this class are confined to the old continent. The South American quadrupeds are on a smaller scale, more feeble and more gentle;

many of them, as the toothless group and the sloths, are of anomalous and less perfect structure than the rest of the animal creation, but the fauna of South America is so local and so peculiar, that the species of five of the terrestrial orders, which are indigenous there, are found nowhere else.

The monkey tribe are in myriads in the forests of tropical America and Brazil, but they never go north of the Isthmus of Darien, nor farther south than the Rio de la Plata. They differ widely from those in the old world, bearing less resemblance to the human race, but they are more gentle and lively, and, notwithstanding their agility, are often a prey to the vulture and puma. Some have no thumb, others have a versable thumb on hands and feet, and the thousands of sapajous have propensile tails, by which they suspend themselves and swing from bough to bough. These inhabitants of the woods are very noisy, especially the argualis, a large ape, whose howling is heard a mile off.

The forests are also inhabited by a family of the marsupial tribe, or animals with pouches, in which they carry their young; they are analogous to those which form the distinguishing feature of the Australian fauna, but of distinct genera and species. All the opossums and the yassacks of this family have thumbs on their hind feet, opposite to the toes, so that they can grasp; they are moreover distinguished from the Australian family by a long prehensile tail, and by greater agility. The numerous tribe of sapajou monkeys, the ant-eaters, the kinkajou, and a species of porcupine, have also grasping tails, a property of many South American animals.

Five genera and 20 species of the toothless quadrupeds are characteristic of this continent, and exclusively confined to South America; they are the sloths, the ai, the armadilloes, chlamyphores, and ant-eaters. The animals of these five genera have very different habits: the sloths, as their name implies, are the most inactive of animals; while the armadillo, in its coat of mail, is in perpetual motion, and in speed can outrun a man. Several species of these animals are nocturnal, and burrow in the earth in the Pampas, Chili, and other places. The chlamyphores are also burrowing animals, peculiar to the province of Cuyo in La Plata, and they have the property of sitting upright. The ant-eater, larger than a Newfoundland dog, with shorter legs, defends itself against the jaguar with its powerful claws; it inhabits the swampy savannahs and damp forests from Columbia to Paraguay, and from the Atlantic to the foot of the Andes; its flesh, like that of some other American animals, has a flavour of musk. The little ant-eater has a prehensile tail, and lives on trees in the tropical forests, feeding on the larvæ of bees, wasps, honey, and ants; another of similar habits lives in Brazil and Guiana. The cat tribe in South America is beautiful and powerful: the puma, the lion of America, is found both in the mountains and the plains, in great numbers; so different are its habits in different places, that in Chili it is timid and flies from a dog; in Peru it is bold, though it rarely attacks a man. The ounce, which inhabits the lower forests, kills Indians even near their huts. The jaguar, a large tiger, very abundant, is so ravenous that it has sprung upon Indians in a canoe; it is one of the few South American animals that cross the Isthmus of Darien, being found in California, on the territory of the Mississippi, and has been seen in Canada.

The vampire is a very large bat, much dreaded by the natives, because it enters their huts at night, and though it seldom attacks human beings, it wounds calves and small animals, which sometimes die from the loss of blood. The other three South American bats are harmless.

The only ruminating animals in South America are the alpacas, vicugnas, llamas, and guanacos; the three first are peculiar to the Andes, the fourth is also in the Pampas, and in all the southern temperate zone to Cape Horn; it is characteristic of the plains of Patagonia, where it is in large herds, and is easily tamed: to these may be added four species of deer. The gnawers of South America are peculiar and varied; more than 40 species of mice are native. The agouti represents our hares on these deserts, and the bizcacha is a burrowing animal, frequent in the pampas of Buenos Ayres. There is only one species of squirrel in the vast forests of South America. The guinea-pig, peccari, and cavies are South American; so is the beautiful chinchilla, the fur of which is so valuable. The only native dogs are a half-reclaimed breed, which the Indians have, and a dumb dog in Brazil.

It is very remarkable that in a country which has the most luxuriant vegetation there should not be one species of hollow-horned ruminants, as the ox, sheep, goat, or antelope; and it is still more extraordinary that the existing animals of South America, which are so nearly allied to the extinct inhabitants of the same soil, should be so inferior in size not only to them, but even to the living quadrupeds of South Africa, which is comparatively a desert. The quantity of vegetation in Britain at any one time exceeds the quantity on an equal area in the interior of Africa ten-fold, yet Mr. Darwin has computed that the weight of 10 of the largest South African quadrupeds is 24 times greater than that of the same number of quadrupeds of South America; for in South America there is no animal the size of a cow, so that there is no relation between the bulk of the species and the vegetation of the countries they inhabit.

The largest animals indigenous in the West Indian Islands are the agouti, the racoon, the houtias, a native of the forests of Cuba; the didelphus carnivora and the kinkajou are common also to the continent: the kinkajou is a solitary instance of a carnivorous animal with a prehensile tail.

AUSTRALIAN QUADRUPEDS.

Australia is not farther separated from the rest of the world by geographical position than by its productions. Its animals are creatures by themselves, of an entirely unusual type; few in species, and still fewer individually, if the vast extent of country be taken into consideration; and there has not been one large animal discovered. There are only 53 species of land quadrupeds in New Holland, and there is not a single example of the ruminating or thick-skinned animals, so useful to man, among them; there are no native horses, oxen, or sheep, yet all these thrive and multiply on the grassy steppes of the country, which seem to be so well suited to them. There are none of the monkey tribe, indeed they could not exist in a country where there is no fruit.

Of the 52 species of indigenous quadrupeds, 40 are found nowhere else, and 43 are marsupial or pouched animals, distinguished from all others by their young being nourished in the pouch till they are mature. Though all the members of this numerous family agree in this circumstance, they are dissimilar in appearance, internal structure, in their teeth and feet, consequently in their habits; two genera live on vegetable food, one set are gnawers and another toothless. The kangaroo and the kangaroo-rat walk on their hind legs, and go by bounds, aided by their strong tail; the rat holds its food in its hands like the squirrel; the opossum walks on all fours; the phalangers live on trees, and swing by their bushy tail, some burrow in the sand; the flying opossum or phalanger. peculiarly an Australian animal, lives on the leaves of the gum-tree; by expanding the skin of its sides it supports itself in the air in its leaps from bough to bough. Several of the genera come out at night only, a characteristic of many Australian animals.

The pouched tribe vary in size from that of a large dog to a mouse; the kangaroos, which are the largest, are easily domesticated, and are used for food by the natives. Some go in large herds in the mountains, others live in the plains; however, they have become scarce near the British colonies, and, with all other native animals, are likely to be exterpated. In Van Diemen's Land they are less persecuted; several species exist there. A wild dog in the woods, whose habits are ferocious, is the largest carnivorous animal in Australia.

The gnawing animals are aquatic and very peculiar, but the toothless animals of New Holland are quite extraordinary; of these there are two genera, the platypus ornithorynchus, or duck-billed mole, and the echidna: they are the link that connects the edentata with the pouched tribe. The duck-billed mole is about 14 inches long, and covered with thick brown fur; its head is similar to that of a quadruped, ending in a bill like that of a duck: it has short furry legs with half-webbed feet, and the hind feet are armed with sharp claws. The burrows it inhabits on the banks of rivers have two entrances, one above, the other below the level of the water, which it seldom leaves, feeding on insects and seeds in the mud.

The echidna is similar in structure to the platypus, but entirely different in external appearance, being covered with quills like the porcupine; it is also a burrowing animal, sleeps during winter, and lives on ants in summer.

A singular analogy exists between Australia and South America in this respect, that the living animals of the two countries are stamped with the type of their ancient geological inhabitants, while in England and elsewhere the difference between the existing and extinct generations of beings is most decided. Australia and South America seem still to retain some of those conditions that were peculiar to the most ancient eras. Thus each tribe of the innumerable families that inhabit the earth, the air, and the waters, has a limited sphere. How wonderful the quantity of life that now is, and the myriads of beings that have appeared and vanished. Dust has returned to dust through a long succession of ages, and has been continually remoulded into new forms of existence—not an atom has been annihilated: the fate of the vital spark that has animated it, with a vividness sometimes approaching to reason, is one of the deep mysteries of Providence.

CHAPTER XXXI.

THE DISTRIBUTION, CONDITION, AND FUTURE PROSPECTS OF THE HUMAN RACE.

More than 860,000,000 of human beings are scattered over the face of the earth, of all nations and kindreds and tongues; and in all stages of civilization, from a high state of moral and intellectual culture, to savages but little above the animals that contend with them for the dominion of the deserts and forests through which they roam. This vast multitude is divided into nations and tribes, differing in external appearance, character, language, and religion. The manner in which they are distributed, the affinities of structure and language by which they are connected, and the effect that climate, food, and customs may have had in modifying their external forms, or their moral and mental powers, are subjects of much more difficulty than the geographical dispersion of the lower classes, inasmuch as the immortal spirit is the chief agent in all that concerns the human race. The progress of the universal mind in past ages, its present state, and the future prospects of humanity, rouse the deep sympathies of our nature for the high but mysterious destiny of the myriads of beings yet to come, who, like ourselves, will be subject for a few brief years to the joys and sorrows of this transient state, and fellow-heirs of eternal life hereafter.

Notwithstanding the extreme diversity, personal and mental, in mankind, anatomists have found that there are no specific differences—that the hideous Esquimaux, the refined and intellectual Circassian, the thick-lipped swarthy Negro, and the fair blue-eyed Scandinavian, are mere varieties of the same species. The human race forms five great classes or families, marked by strong distinctive characters. Many nations are included in each, distinguished from one another by different languages, manners, and mental qualities, yet bearing such a resemblance in structure and physiognomy as to justify a classification apparently anomalous.

The Circassian group of nations, which includes the handsomest and most intellectual portion of mankind, inhabit all Europe, except Lapland, Finland, and Hungary; they occupy North Africa as far as the 20th parallel of north latitude, Arabia, Asia Minor, Persia, the Himalaya to the Brahmapootra, all India between these mountains and the ocean, and the United States of North America. These nations are remarkable for a beautifully-shaped small head, regular features, fine hair, and symmetrical form. The Greeks, Georgians, and Circassians are models of perfection in form, especially the last, assumed as the type of this class of mankind; of which it is evident that colour is not a characteristic, since they are of all shades, from the fair and florid to the clear dark brown and almost black. This family of nations has always been, and still is, the most civilized portion of the human race. The inhabitants of Hindostan, the Egyptians, Arabians, Greeks, and Romans, were in ancient times what the European nations are now. The cause of this remarkable development of mental power is no doubt natural disposition, for the difference in the capabilities of nations seems to be as great as that of individuals. The origin of spontaneous civilization and superiority may generally be traced to the talent of some master-spirit gaining an ascendancy over his countrymen. Natural causes have also combined with mental—mildness of climate, fertility of soil; rivers and inland seas, by affording facility of intercourse, favoured enterprise and commerce; and the double-river systems in Asia brought distant nations together, and softened those hostile antipathies which separate people, multiply languages, and reduce all to barbarism. The genius of this family of nations has led them to profit by these natural advantages, whereas the American Indians are at this day wandering as barbarous hordes in one of the finest countries in the world. An original similarity or even identity of many of the spoken languages, may be adverted to as facilitating communication and mental improvement among the Circassian class in very ancient times.

The Mongol-Tartar family forms the second group of nations. They occupy all Asia north of the Persian table-land and of the Himalaya, the whole of eastern Asia from the Brahmapootra to Behring's

Straits, together with the Arctic regions of North America south to Labrador. This family includes the Tourkomans, Mongol and Tartar tribes, the Chinese, Indo-Chinese, Japanese, and Esquimaux; and the Hungarians in the very heart of Europe. These nations are distinguished by broad skulls and high cheek-bones, with small black eyes obliquely set, long black hair, and a yellow or sallow olive complexion; some are good looking, and many are well made. A portion of this family is capable of high culture, especially the Chinese, the most civilized nation of eastern Asia, although they never have attained the excellence of the Caucasian group, probably from their exclusive social system, which has separated them from the rest of mankind and kept them stationary for ages; the peculiarity and difficulty of their language have also tended to insulate them. The Kalmuks, who lead a pastoral, wandering life, on the steppes of Central Asia, and the Esquimaux, have wider domains than any other of this set of nations. The Kalmuks are rather a handsome people, and, like all who lead a savage life, have acute senses of seeing and hearing. The inhabitants of Finland and Lapland are nearly allied to the Esquimaux, who occupy all the high latitudes of both continents—a diminutive race, equally ugly in face and form.

Malayan nations occupy the Indian Archipelago, New Zealand, Chatham Island, the Society group, and several other of the Polynesian islands, together with the Philippines and Formosa. They are very dark, with lank coarse black hair, flat face, and obliquely set eyes. Endowed with great activity and ingenuity, they are mild and gentle, and far advanced in the arts of social life, in some places; in others, ferocious and revengeful, daring and predatory: and, from their maritime position and skill, they are a migratory race. Several branches of this class of nations had a very early indigenous civilization, with an original literature in peculiar characters of their own.

The Ethiopian nations are widely dispersed; they occupy all Africa south of the Great Desert, half of Madagascar, the continent of Australia, Mindanao, Gilolo, the high lands of Borneo, Sumbawa, Timor, and New Ireland. The distinguishing characters of this group are a black complexion, black woolly or frizzled hair, thick lips, projecting jaws, high cheek-bones, and large prominent eyes. A great variety, however, exists in this jetty race. Some are handsome both in face and figure, especially in Ethiopia; and even in Western Africa, where the negro tribes live, there are groups in which the distinctive characters are less exaggerated. This great family has not yet attained a high place among the nations, though by no means incapable of cultivation; and part of Ethiopia appears to have made considerable advances in civilization in very ancient times. But the formidable deserts, so extensive in some parts of the continent, and the unwholesome climate in others, have cut off the intercourse with civilized nations; and, unfortunately, the infamous traffic in slaves, to the disgrace of Christianity, has made the nations of tropical Africa more barbarous than they were before; while, on the contrary, the Foulahs and other tribes, who were converts to Mohammedanism 400 years ago, have now large commercial towns, cultivated grounds, and schools. The Australians and Papuans, who inhabit the Eastern islands mentioned, are the most degraded of this dark race, and indeed of all mankind.

The American race, who occupy the whole of that continent from 62° N. lat. to the Straits of Magellan, are almost all of a reddish brown or copper colour, with long black hair, deep-set black eyes, aquiline nose, and often of handsome slender forms. In North America they live by hunting, are averse to agriculture, slow in acquiring knowledge, but extremely acute, brave, and fond of war; and though revengeful, are capable of generosity and gratitude. In South America many are half civilized, but a greater number are still in a state of utter barbarism. In a family so widely scattered, great diversity of character prevails; yet throughout the whole there is a similarity of manners and habits, which has resisted all the effects of time and climate.

Each of these five groups of nations, spread over vast regions, is accounted one family; and if they are so by physical structure, they are still more so by language, which expresses the universal mind of a people, modified by external circumstances, of which none have a greater influence than the geographical features of the country they inhabit, and that influence is deepest in the early stages of society. The remnants of ancient poetry in the south of Scotland partake of the gentle and pastoral character of the country; while Celtic verse, and even the spoken language of the Highlander, are full of the poetical images of war and stern mountain scenery. As civilization advances, and man becomes more intellectual, the language keeps pace in the progress. New words and new expressions are added, as new ideas occur and new things are invented, till at last language itself becomes a study, is refined, and perfected by the introduction of general terms. The art of printing perpetuates a tongue, and great authors immortalize it; yet language is ever changing to a certain degree, though it never loses traces of its origin. Chaucer and Spenser have become obscure; Shakspeare requires a glossary for the modern reader; and in the few years that the United States of America have existed as an independent nation the speech has deviated from the mother tongue. When a nation degenerates, it is split by jealousy and war into tribes, each of which in process of time acquires a peculiar idiom, and thus the number of dialects is increased, though they still retain a similarity; whereas when masses of mankind are united into great political bodies, their languages by degrees assimilate to one common tongue, which retains traces of all to the latest ages. The form of the dialects now spoken by some savage tribes, as the North American Indians, bears the marks of a once higher state of civilization.

More than 2000 languages are spoken, but few are independent; some are connected by words having the same meaning, some by grammatical structure, others by both; indeed the permanency of language is so great, that neither ages of conquest nor mixing with other nations have obliterated the native idiom of a people. The French, Spanish, and German retain traces of the common language spoken before the Roman conquest, and the Celtic tongue still exists in the British Islands.

By a comparison of their dialects, nations far apart, and differing in every other respect, are discovered to have sprung from a common, though remote origin. Thus all the numerous languages spoken by the American Indians, or red men, are similar in grammatical structure: an intimate analogy exists in the languages of the Esquimaux nations, who inhabit the Arctic regions of both continents. Dialects of one tongue are spoken throughout North Africa, as far south as the oasis of Siwah on the east and the Canary Islands on the west. Another group of cognate idioms is common to the inhabitants of equatorial Africa; while all the southern part of the continent is inhabited by people whose languages are connected. The monosyllabic speech of the Chinese and Indo-Chinese shows that they are the same people, and all the insular nations of the Pacific derived their dialects from some tribes on the continent of India and the Indian Archipelago.

The Persian, Arabic, Greek, Latin, German, and

Celtic tongues are connected by grammatical structure, and words expressive of the same objects and feelings with the Sanscrit, or sacred language of India; consequently the nations inhabiting the British Islands and these extensive districts of the Continent must have had the same origin.

The two methods of classing mankind that have been mentioned do not perfectly agree, nor does either of them include the whole, but an approximation is all that can be attained in so complicated a subject.

It is no difficult matter to see how changes may occur in speech, but no circumstance in the natural world is more inexplicable than the diversity of form and colour in the human race. It had already begun in the Antediluvian world, for "there were giants in the land in those days." No direct mention is made of colour at that time, unless the mark set upon Cain, "lest any one finding him should kill him," may allude to it. Perhaps, also, it may be inferred that black people dwelt in Ethiopia, or the land of Cush, which means black in the Hebrew tongue. At all events, the difference now existing must have arisen after the flood, consequently all must have originated with Noah, whose wife, or the wives of his sons, may have been of different colours for ought we know.

Many instances have occurred in modern times of albinos and red-haired individuals having been born of black parents, and these have transmitted their peculiarities to their descendants for several generations; but it may be doubted whether pure-blooded white people have ever had perfectly black offspring. The varieties are much more likely to have arisen from the effects of climate, food, customs, and civilization upon migratory groups of mankind, and of such a few instances have occurred in historical times, limited, however, to small numbers and particular spots; but the great mass of nations had received their distinctive characters at a very early period. The permanency of type is one of the most striking circumstances, and shows the immense length of time necessary to produce a change in national structure and colour. A nation of Ethiopians existed 3450 years ago, which emigrated from a remote country and settled near Egypt, and there must have been black people before the age of Solomon, otherwise he would not have alluded to colour even poetically. Besides, the national appearance of the Ethiopians, Persians, and Jews has not varied for more than 3000 years, as appears from the ancient Egyptian paintings in the tomb of Rhameses the Great, discovered at Thebes by Belzoni, in which the countenance of the modern Ethiopian and Persian can be readily recognised, and the Jewish features and colour are identical with those of the Israelites daily met with in London. As there is no instance of a new variety of mankind having been established as a nation since the Christian era, there must either have been a greater energy in the causes of change before that era, or, brief as man's span on earth has been, a wrong estimate of time antecedent to the Christian period must have made it shorter.

Darkness of complexion has been attributed to the sun's power from the age of Solomon to this day. "Look not upon me, because I am black, because the sun hath looked upon me:" and there cannot be a doubt, that to a certain degree the opinion is well founded. The invisible rays in the solar beams, which change vegetable colour, and have been employed with such remarkable effect in the Daguerreotype, act upon every substance on which they fall, producing mysterious and wonderful changes in their molecular state, man not excepted.

Other causes must have been combined to occasion all the varieties we now see, otherwise every nation between the tropics would be of the same hue; whereas the sooty negro inhabits equatorial Africa, the red man equinoctial America, and both are mixed with fairer tribes. In Asia, the Rohillas, a fair race, inhabit the plains south of the Ganges; the Bengalee and the mountaineers of Nepaul are dark, and the Mahrattas are yellow. Even supposing that diversity of colour is owing to the sun's rays only, it is scarcely possible to attribute the thick lips, the woolly hair, and the entire difference of form, extending even to the very bones and skull, to anything but a variety of concurring circumstances, not omitting the invisible influence of electricity, which pervades every part of the earth and air, and possibly terrestrial magnetism.

The flexibility of man's constitution enables him

to live in every climate from the equator to the ever frozen coasts of Nova Zembla and Spitzbergen, and that chiefly by his capability of bearing the extremest changes of temperature and diet, which are probably the principal causes of the variety in his form. It has already been mentioned that oxygen is inhaled with the atmospheric air, and also taken in by the pores in the skin; part of it combines chemically with the carbon of the food, and is expired in the form of carbonic-acid gas and water; that chemical action is the cause of vital force and heat in man and animals. The quantity of food must be in exact proportion to the quantity of oxygen inhaled, otherwise disease and loss of strength would follow. Since cold air is incessantly carrying off warmth from the skin, more exercise is requisite in winter than in summer, in cold climates than in warm; consequently more carbon is necessary in the former than in the latter, in order to maintain the chemical action that generates heat, and to ward off the destructive effects of the oxygen, which incessantly strives to consume the body. Animal food, wine, and spirits contain many times more carbon than fruit and vegetables, therefore animal food is much more necessary in a cold than in a hot climate. The Esquimaux, who lives by the chace, and eats 10 or 12 pounds weight of meat and fat in 24 hours, finds it not more than enough to keep up his strength and animal heat; while the indolent inhabitant of Bengal is sufficiently supplied with both by his rice diet. Clothing and warmth make the necessity for exercise and food much less, by diminishing the waste of animal heat. Hunger and cold united soon consume the body, because it loses its power of resisting the action of the oxygen, which consumes part of our substance when food is wanting. Hence nations inhabiting warm climates have no great merit in being abstemious, nor are those committing an excess who live more freely in the colder countries. The arrangement of Divine Wisdom is to be admired as much in this as in all other things, for if man had only been capable of living on vegetable food, he never could have had a permanent residence beyond the latitude where corn ripens. The Esquimaux and all the inhabitants of the very high latitudes of both continents live entirely on fish and animal food.

A nation or tribe driven by war, or any other cause, from a warm to a cold country, or the contrary, would be forced to change their food both as to quantity and quality, which in the lapse of ages might produce an alteration in the external form and internal structure. The probability is still greater, if the entire change that a few years produces in the matter of the human frame be considered. At every instant during life, with every motion, voluntary and involuntary, with every thought and exercise of the brain, a portion of our substance becomes dead, separates from the living part, combines with some of the inhaled oxygen, and is removed. By this process it is supposed that the whole body is renewed every seven years: individuality, therefore, depends on the spirit, which retains its identity during all the changes of its earthly house, and sometimes even acts independently of it. When sleep is restoring exhausted nature, the spirit is often awake and active, crowding the events of years into a few seconds, and, by its unconsciousness of time, anticipates eternity. Every change of food, climate, and mental excitement must have their influence on the reproduction of the mortal frame; and thus a thousand causes may co-operate to alter whole races of mankind placed under new circumstances, time being granted.

The difference between the effects of manual labour and the efforts of the brain appears in the intellectual countenance of the educated man, compared with that of the peasant, though he also is occasionally stamped with nature's own nobility. The most savage people are also the ugliest. Their countenance is deformed by violent unsubdued passions, anxiety, and suffering. Deep sensibility gives a beautiful and varied expression, but every strong emotion is unfavourable to perfect regularity of feature; and of that the ancient Greeks were well aware when they gave that calmness of expression and repose to their unrivalled statues. The refining effects of high culture, and, above all, the Christian religion, by subduing the evil passions, and encouraging the good, are more than anything calculated to improve even the external appearance. The countenance, though perhaps of less regular form, becomes expressive of the amiable and benevolent feelings of the heart, the most captivating and lasting of all beauty.

Thus an infinite assemblage of causes may be assigned as having produced the endless varieties in the human race; but the fact remains an inscrutable mystery not to be explained, more than why twin-brothers are not exactly alike. But amidst all the physical vicissitudes man has undergone, the species remains permanent; and let those who think that the difference in the species of animals and vegetables arises from diversity of conditions, consider that no circumstances whatever can degrade the form of man to that of the monkey, or elevate the monkey to the form of man.

Animals and vegetables, being the sources of man's sustenance, have had the chief influence on his destiny and location, and have induced him to settle in those parts of the world where he could procure them in greatest abundance. Wherever the chace or the spontaneous productions of the earth supply him with food, he is completely savage, and only a degree further advanced where he plants the palm and banana; where grain is the principal food, industry and intelligence are most perfectly developed, as in the temperate zone. On that account, the centres of civilization have generally been determined, not by hot, but genial climate, fertile soil, by the vicinity of the sea-coast or great rivers, affording the means of fishing and transport, which last has been one of the chief causes of the superiority of Europe and southern Asia. The mineral treasures of the earth have been the means of assembling great masses of men in Siberia and the table-land of the Andes, and have given rise to many great cities, both in England and North America. Nations inhabiting elevated table-lands and high ungenial latitudes have been driven there by war, or obliged to wander from countries where the population exceeded the means of living—a cause to which both language and tradition bear testimony. The belief in a future state, so universal, and shown by respect for the dead, has no doubt been transmitted from nation to nation. The American Indians, driven from their hunting-grounds, still make pilgrimages to the tombs of their fathers; and these tribes alone, of all uncivilized mankind, worship the Great Spirit as the invisible God and Father of all —a degree of abstract refinement which could hardly have sprung up spontaneously among a rude people, and must have been transmitted from races who held the Jewish faith.

The influence of external circumstances on man is not greater than his influence on the material world. It is true, he cannot create power; but he dexterously avails himself of the powers of nature to subdue nature. Air, fire, water, steam, gravitation, his own muscular strength and that of animals, have been the instruments by which he has converted the desert into a garden, drained marches, turned the course of rivers, cut canals, made roads, cleared away forests in one country, and planted them in another. By these works he has altered the climate, changed the course of local winds, increased or diminished the quantity of rain, and softened the rigour of the seasons. In the time of Strabo, the cold in France was so intense, that it was thought impossible to ripen grapes north of the Cevennes; and the Rhine and Danube were every winter covered with ice thick enough to bear any weight. Man's influence on vegetation has been immense, but the most important changes were produced in the antediluvian ages of the world. Cain was a tiller of the ground. The olive, the vine, and the fig-tree have been cultivated time immemorial: wheat, rice, and barley have been so long in an artificial state that their origin is unknown: even maize, which is a Mexican plant, was in use among the American tribes before the Spanish conquest; and tobacco was already used by them to allay the pangs of hunger, to which those who depend upon the chace for food must be exposed. Most of the ordinary culinary vegetables have been known for ages; and it is singular that in these days, when our gardens are adorned with innumerable native plants in a cultivated state, no new grain, vegetable, or fruit has been reclaimed: the old have been produced in infinite variety, and many brought from foreign countries; yet there must exist many plants capable of cultivation, as unpromising in their wild state as the turnip or carrot.

Some families of plants are more susceptible of improvement than others, and, like man himself, can bear almost any climate. One kind of wheat grows to 62° N. lat.; rye and barley succeed still farther north; and few countries are absolutely without grass. The cruciform tribe abounds in useful plants; indeed, that family, together with the solanum, the papilionaceous, and umbelliferous tribes, furnish most of our vegetables. Many plants, like animals, are of but one colour in their wild state, and their blossoms are single. Art has introduced that variety we now see in the same species; and by changing the anthers of the wild flower into petals, has produced double blossoms: by art, too, many plants of warm countries have been naturalized in colder. Few useful plants have beautiful blossoms; but if utility were the only object, of what pleasure should we be deprived? Refinement is not altogether wanting in the inmates of a cottage covered with roses and honeysuckle; and the little garden, cultivated amidst a life of toil, tells of a peaceful home.

Among the objects which tend to the improvement of our race, the flower-garden and the park, adorned with native and foreign trees, have no small share, they are the greatest ornaments of the British Islands; and the love of a country life, which is so strong a passion, is chiefly owing to the law of primogeniture, by which the head of a family is secured in the possession and transmission of his undivided estate, and therefore each generation takes a pride and pleasure in adorning the home of their forefathers.

Animals yield more readily to man's influence than vegetables; but certain classes have a greater flexibility of disposition and structure than others.

Those only are capable of being perfectly reclaimed that have a natural tendency for it, without which man's endeavours would be unavailing. This predisposition is greatest in animals that are gregarious and follow a leader, which elephants, dogs, horses, and cattle do in their wild state; but even among these some species are refractory, as the buffalo, which can only be regarded as half reclaimed. The canine tribe, on the contrary, are capable of the greatest attachment; not the dog only, man's faithful companion, but even the wolf, and especially the hyæna, generally believed to be so ferocious. After an absence of many months, a hyæna recognised the voice of a friend of the author before he came in sight, and on seeing him it showed the greatest joy, lay down like a dog and licked his hands. He had been kind to it on the voyage from India, and no animal forgets kindness, which is the surest way of reclaiming them. There cannot be a greater mistake than the harsh and cruel means by which dogs and horses are too commonly trained; but it is long before man learns that his power is mental, and that it is his intellect alone that has given him dominion over the earth and its inhabitants, many of which far surpass him in physical strength. The useful animals were reclaimed by the early inhabitants of Asia, and little has been left for modern nations but the improvement of the species, and in that they have been very successful. The variety of horses, dogs, oxen, and sheep is beyond number. The form, colour, and even the disposition, may be materially altered, and the habits engrafted are transmitted to the offspring, as instinctive properties independent of education. Domestic fowls go in flocks in their native woods when wild. There are, however, instances of solitary birds being tamed to an extraordinary degree, as the raven, one of the most sagacious.

Man's necessities and pleasure have been the cause of great changes in the animal creation, but his destructive propensity of still greater. Animals are intended for our use, and field-sports are advantageous by encouraging a daring and active spirit in young men; but the utter destruction of some races, in order to protect those destined for his amusement, is too selfish. Animals soon acquire an instinctive dread of man, which becomes hereditary. In newly discovered uninhabited countries, birds and beasts are so tame as to allow themselves to be taken. Whales scarcely got out of the way of the ships that first navigated the Arctic Ocean, but they now have a dread of the common enemy. Many land animals and birds are vanishing before the advance of civilization. Sea-fowl and birds of passage are not likely to be extinguished. The inaccessible cliffs of the Himalaya and the Andes will afford a refuge to the eagle and the condor; but the time will come when the mighty forests of the Amazons and Orinoco will disappear with the myriads of their joyous inhabitants. The lion, the tiger, and the elephant will be known only by ancient records. Man, the lord of the creation, will extirpate the noble creatures of the earth, but he himself will ever be the slave of the canker-worm and the fly. Cultivation may lessen the scourge of the insect tribe, but God's great army will ever from time to time appear suddenly, no one knows from whence; and the locust will come from the desert, and destroy the fairest prospects of the harvest.

Though the unreclaimed portion of the animal creation is falling before the progress of improvement, yet man has been both the voluntary and the involuntary cause of the introduction of new animals and plants into countries in which they were not native. The Spanish conquerors little thought that the descendants of the horses and cattle they allowed to run wild would resume the original character of their species, and roam in hundreds of thousands over the savannahs of South America. Wherever man is, civilized or savage, there also is the dog; but he too has in some places resumed his native state and habits, and hunts in packs. Domestic animals, grain, fruit, vegetables, and the weeds that grow with them, have been conveyed by colonists to all settlements. Birds and insects follow certain plants into countries in which they were never seen before. Even the inhabitants of the waters change their abode in consequence of the influence of man. Fish, natives of the rivers on the coast of the Mexican Gulf, have migrated by the canals to the heart of North America; and the mytilus polymorphus, a shell-fish brought to the London docks in the timbers of ships from the brackish waters of the Black

Sea and its tributary streams, has spread into the interior of England by the Croydon and other canals.

The influence of man on man is a power of the highest order, far surpassing that which he possesses over inanimate or animal nature; and at no time did the mental superiority of the cultivated races produce such changes as they do at present. In civilized society, the number of people in the course of time exceeds the means of sustenance, which compels some to emigrate; others are induced by a spirit of enterprise to go to new countries, some for the love of gain, others to fly from oppression.

The discovery of the new world opened a wide field for emigration. Spain and Portugal, the first to avail themselves of it, acquired dominion over some of the finest parts of South America, which they have maintained, till lately a change of times has rendered their colonies independent states. Liberal opinions have spread into the interior of the continent in proportion to the facility of communication with the cities on the coasts, from whence European ideas are disseminated. Of this Venezuela is an instance, where civilization and prosperity have advanced more rapidly than in the southern parts of Columbia, where the Andes are higher, and the distance from the Atlantic greater. Civilization has been impeded in many of the smaller states by war and those broils inevitable among people unaccustomed to free institutions; and Brazil would have been farther advanced but for slavery, that stain on the human race, which corrupts the master as much as it debases the slave.

Some of the native South American tribes have spontaneously made considerable progress in civilization in modern times; others have benefited by the Spanish and Portuguese colonists; and many tribes have been brought into subjection by the Jesuits, who have instructed them in some of the arts of social life: but these Indians are not more religious than their neighbours; and, from restraint, they have lost vigour of character without improving in intellect, so that they are now either stationary or retrograde. But extensive regions are still the abode of men in the lowest state of barbarism; almost all those inhabiting the silvas of the Orinoco, Amazons, and Uruguay are cannibals.

The arrival of the colonists in North America sealed the fate of the red man. The inhabitants of the Union, too late awakened to the just claims of the ancient proprietors of the land, have recently, but vainly, attempted to save the remnant. The white man, like an irresistible torrent, has already reached the centre of the continent; and the native tribes now retreat towards the far west, and will continue to retreat till the Pacific Ocean arrests them, and the animals on their hunting-grounds are exterminated. The almost universal dislike the Indian has shown for the arts of peace has been one of the principal causes of his decline, although the Cherokee tribe, which has lately removed to the west of the Mississippi, is a remarkable exception:

the greater number of them are industrious planters or mechanics; they have a republican government, and publish a newspaper in their own language, in a character lately invented by one of that nation.

No part of the world has been the scene of greater iniquity than the West Indian Islands, and that perpetrated by the most enlightened nations of Europe. The native race has long been swept away by the stranger, and a new people, cruelly torn from their homes, have been made the slaves of hard taskmasters. If the odious participation in this guilt has been a stain on the British name, the abolition of slavery by the universal acclamation of the nation will ever form one of the brightest pages in their history, so full of glory; nor will it be the less so that justice was combined with mercy, by the millions granted to indemnify the proprietors. It is deeply to be lamented that our brethren on the other side of the Atlantic have not followed the example of their fatherland; but in limited monarchies the voice of the people is listened to, while republican governments are more apt to become its slave. The northern States have nobly declared every man free who sets his foot on their territory; and the time will come when the southern States will sacrifice interest to justice and mercy.

It seems to be the design of Providence to supplant the savage by civilized man in the continent of Australia as well as in North America, though every effort has been made to prevent the extinction of the natives. Most of the tribes in that continent are as low in the rank of mankind as the cannibal Fuegians whom Captain Fitzroy so generously but ineffectually attempted to tame. Some of the New Hollanders are faithful servants for a time; but they almost always return to their former habits, though truly miserable in a country where the means of existence are so scanty. Animals and birds are very scarce; and there is no fruit or vegetable for the sustenance of man.

Slavery has been a greater impediment to the improvement of the nations of Africa than even the physical disadvantages of the country, the great arid deserts and unwholesome coasts. A spontaneous civilization has arisen in various parts of southern and tropical Africa, in which there has been considerable progress in agriculture and commerce; but civilized man has been a scourge on the Atlantic coast, which has extended its influence into the heart of the continent, by the encouragement it has given to warfare among the natives for the capture of slaves, and by the introduction of European vices, unredeemed by Christian virtues. Now that France and England have united in the suppression of this odious traffic, some hope may be entertained that their colonies may be beneficial to the natives, and that other nations may follow their example, in which, however, they have been anticipated by three Mohammedan sovereigns. The Sultan has abolished the slave-market in Constantinople; Ibrahim Pasha, on his return from France and England, gave freedom to his bondsmen in Egypt;

and the Bey of Tunis has abolished slavery in his dominions.

The French are zealous in improving the people in Algiers; but the constant state of warfare in which they have been involved ever since their conquest, must render their success in civilizing the natives at least remote. And the inhabitants of those extensive and magnificent countries that have long been colonized by the Dutch, have made but little progress under their rule.

The British colony at the Cape of Good Hope has had considerable influence on the neighbouring rude nations, who begin to adopt more civilized habits. When Mr. Somerville visited Latakoo, the natives were scantily covered with skins, and they saw horses for the first time. Dr. Smith, who visited them twenty years afterwards, found the chief men mounted on horseback, wearing hats made of rushes, and an attempt to imitate European dress.

Colonization has nowhere produced such happy effects as among the amiable and cultivated inhabitants of India, who are sensible of the benefits they derive from the impartial administration of just and equal laws, the foundation of schools and colleges, and the extension of commerce.

All the causes of emigration have operated by turns on the inhabitants of Britain, and various circumstances have concurred to make their colonies permanent. In North America, that which not many years ago was a British colony has become a great independent nation, occupying a large portion of the continent. The Australian continent will in after ages be peopled by British nations, and will become a centre of civilization which will extend its influence to the uttermost islands of the Pacific. These splendid islands, possessing every advantage of climate and soil, with a population in many parts far advanced in the arts of civilized life, industry, and commerce, though in others savage, will in time come in for a share of the general improvement. The success that has attended the noble and unaided efforts of Mr. Brooke in Borneo, shows how much the influence of an active mind can effect.

The colonies on the continent of India are already centres from which the culture of Europe is spreading over the East.

Commerce has no less influence on mankind than colonization, with which it is intimately connected; and the narrow limits of the British Islands have rendered it necessary for its inhabitants to exert their industry for their well-being. The riches of our mines in coal and metals, which produce a yearly income of 24,000,000*l*. sterling, is a principal cause of our manufacturing and commercial wealth; but even with these natural advantages, more is due not only to the talents and enterprise, but to our high character for faith and honour.

Every country has its peculiar productions, and by an unrestrained interchange of the gifts of Providence the condition of all is improved. The exclusive jealousy with which commerce has hitherto been fettered, shows the length of time that is necessary to wear out the effects of those selfish passions which separated nations when they were yet barbarous. It required a high degree of cultivation to break down those barriers consecrated by their antiquity, and the accomplishment of this important change evinces the rate at which the present age is advancing.

A new era in the history of the world began when China was opened to European intercourse; but many years must pass before European influence can penetrate that vast empire, and eradicate those illiberal prejudices by which it has so long been governed.

Two important triumphs yet remain to be achieved by the science and energy of man over physical difficulties, namely, the junction of the Pacific and Atlantic oceans at the Isthmus of Central America, and the union of the Red Sea with the Mediterranean at Suez: when these are accomplished, the expectation of Columbus will be realized—of a passage to the East Indies by the Atlantic; then Alexandria, Venice, and Southern Europe will regain, at least in part, the mercantile position which they lost by the discovery of Vasco de Gama.

The advantages of colonization and commerce to the less civilized part of the world are incalculable, as well as to those at home, not only by furnishing an exchange for manufactures, important as it is, but by the immense accession of knowledge of the earth and its inhabitants that has been thus attained.

The history of former ages exhibits nothing to be compared with the mental activity of the present. Steam, which annihilates time and space, fills mankind with schemes for advantage or defence; but however mercenary the motives for enterprise may be, it is instrumental in bringing nations together, and uniting them in mutual bonds of friendship. The facility of communication is rapidly assimilating national character. Society in most of the capitals is formed on the same model; individuality is only met with in the provinces, and every well educated person now speaks more than one of the modern languages.

At no period has science been so extensively and so successfully cultivated; the collective wisdom and experience of Europe and the United States of America is now brought to bear on subjects of the highest importance in annual scientific meetings, where the common pursuit of truth is as beneficial to the moral as to the intellectual character, and the noble objects of investigation are no longer confined to the philosophic few, but are becoming widely diffused among all ranks in civilized nations, and the most enlightened governments have given their support to measures that could not have been otherwise accomplished. Simultaneous observations are made in numerous places, in both hemispheres, on electricity, magnetism, on the tides and currents of the air and the ocean, and those mysterious vicissitudes of temperature and moisture which bless the labours of the husbandman one year and blight them in another.

The places of the nebulæ and fixed stars, and their motions, are known with unexampled precision, and the most refined analyses embrace the most varied objects. In the far heavens, from unaccountable disturbances in the motions of Uranus, an unknown and unseen body was declared to be revolving on the utmost verge of the solar system : it was found in the very spot pointed out by analysis; and on earth, though hundreds of miles apart, the invisible messenger, electricity, instantaneously conveys the thoughts of the invisible spirit of man to man —results of science sublimely transcendental.

The attempt would be vain to enumerate the improvements in machinery and mechanics; to follow the rapid course of discovery through the complicated mazes of magnetism and electricity, the action of the electric current on the polarized sun-beam, one of the most beautiful of modern discoveries, leading to relations hitherto unsuspected between that power and the complex assemblage of visible and invisible influences in solar light, by one of which nature has recently been made to paint her own likeness. It is equally impossible to convey an idea of the rapid succession of the varied and curious results of chemistry, and its application to physiology and agriculture; moreover, distinguished works have lately been published at home and abroad on the science of mind, which has been so successfully cultivated in our own country. Geography has assumed a new character by that unwearied search for accurate knowledge and truth that marks the present age, and physical geography is altogether a modern science.

The spirit of nautical and geographical discovery, begun in the fifteenth century, by those illustrious navigators who had a new world to discover, is at this day as energetic as ever, though the results are necessarily less brilliant. Neither the long gloomy night of a polar winter nor the dangers of the ice and the storm deter our gallant seamen from seeking a better acquaintance with "this ball of earth," even under its most frowning aspect, and that for honour, which they are as eager to seek even in the cannon's mouth. Nor have other nations of Europe or America been without their share in these bold adventures. The scorching sun and deadly swamps of the tropics as little prevent the traveller from collecting the animals and plants of the present creation, or the geologist from investigating those of ages long gone by. Man daily vindicates his birthright as lord of the creation, and compels every land and sea to contribute to his knowledge.

The most distinguished modern travellers, following the example of Baron Humboldt, the patriarch of physical geography, take a more extended view of the subject than the earth and its animal and vegetable inhabitants afford, and include in their researches the past and present condition of man, the origin, manners, and languages of existing nations, and the monuments of those that have been. Geography has had its dark ages, during which, the situation of many great cities, and spots of celebrity in history, sacred and profane, had been entirely lost sight of; which are now discovered by the learning and assiduity of the modern traveller. Of this, Italy, Egypt, the Holy Land, Asia Minor, Arabia, and the basin of the Euphrates and Tigris, with the adjacent mountains of Persia, are remarkable instances, not to mention the vast region of the East. In many parts of the world the ruins of cities, of extraordinary magnitude and workmanship, show that there are wide regions of whose original inhabitants we know nothing. The Andes of Mexico and Peru have remains of civilized nations before the Incas; Mr. Stephens has found in the woods of Central America the ruins of great cities, adorned with sculpture and pictorial writings, which show that a race far advanced had once cultivated the soil where these entangled forests now grow. Picture-writings have been discovered by Mr. Schomburgk on rocks in Guiana, spread over an extent of 350,000 square miles, similar to those found in the United States and in Siberia. Magnificent buildings still exist, in good preservation, all over Eastern Asia, and many in a ruinous state belong to a period far beyond written record.

Ancient literature has furnished a subject of still more interesting research, which shows that the mind of man is essentially the same under very different circumstances; every nation far advanced in civilization has had its age of poetry, the drama, romance, and philosophy, each stamped with the character of the people and times, and still more with their religious belief. Our profound Oriental scholars have made known to Europeans the refined Sanscrit literature of Hindostan, its schools of philosophy and astronomy, its dramatic writings and poetry, which are original and beautiful.

The riches of Chinese literature and their valuable geography were introduced into Europe by the French Jesuits of the last century, and perfected by the French philosophers of the present; to that nation we also owe our knowledge of the letters and poetry of ancient Persia; and from the time that Dr. Young deciphered the inscriptions on the Rosetta Stone, Egyptian hieroglyphics and picture-writing have been studied by the learned, and we have reason to expect much new information from Professor Lipsius, of Berlin. The Germans indeed have left no subject of ancient literature unexplored, even to the language spoken at Babylon and Nineveh.

The press has overflowed with an unprecedented quantity of literature, some of standard merit, and much more that is ephemeral, suited to all ranks and on every subject; and with the aim, in our own country at least, to improve the people and to advocate the cause of virtue. All this mental energy is but an effect of those laws which regulate human affairs, and include in their generality the various changes that tend to improve the condition of man.

The fine arts do not keep pace with science, though they have not been altogether left behind. Painting, like poetry, must come spontaneously, because a feeling for it depends upon innate sympathies in the human breast. Nothing external could affect us unless there were corresponding ideas within; and poetically constituted minds of the highest organization are most deeply impressed with whatever is excellent. All are not gifted with a strong perception of the beautiful, just as some persons cannot see certain colours or hear certain sounds. Those elevated sentiments which constitute genius are given to few; yet something akin, though inferior in degree, exists in most men. Consequently, though culture may not inspire genius, it cherishes and calls forth the natural perception of what is good and beautiful, and by that means improves the tone of the national mind, and forms a counterpoise to the all-absorbing useful and commercial.

Historical painting is successfully cultivated both in France and Germany. The Germans have modelled their school on the true style of the ancient masters. They have not, indeed, attained their richness of colouring, but many of their designs are poetry embodied; and French artists, following in the same steps, have produced historical works of extraordinary merit. Pictures of the *genre* and scenes of domestic life have been painted with much expression and beauty by our own artists; and British landscapes are not mere portraits of nature, but pictures of high poetical feeling; and the perfection of their composition has been acknowledged all over Europe by the popularity of the engravings that illustrate many of our modern books. The encouragement given to this branch of art at home may be ascribed to the taste for a country life so general in England. Water-colour painting, which is entirely of British growth, has now become a favourite style in every country, and is brought to the highest perfection in our own.

The Italians have had the merit of restoring sculpture to the pure style which it had lost; and that gifted people have produced some of the noblest specimens of modern art. The greatest genius of his time left the snows of the far North to spend his best days in Rome, the head-quarters of the art; and our own sculptors of the most eminent talents have established themselves in Rome, where they find a more congenial spirit than in their own country, where the compositions of Flaxman were not appreciated till they had become the admiration of Europe.

The opera, one of the most refined of theatrical amusements in every capital city in Europe, shows the power and excellence of Italian melody, which has been transmitted from age to age by a continued succession of great composers. German music, partaking of the learned character of the nation, is rich in original harmony, which requires a cultivated taste to understand and appreciate.

Italy is the only country that has had two poetical eras of the highest order; and great as the Latin period was, that of Dante was more original and sublime. The Germans, so eminent in every branch of literature, have been also great as poets: the power of Göthe's genius will render his poems as permanent as the language. France is, as it long has been, the abode of the Comic Muse; and although that nation can claim great poets of a more serious cast, yet the language and the habits of the people are more suited to the gay than the grave style. Though the British may have been inferior to other nations in some of the fine arts, yet poetry, immeasurably the greatest and most noble, redeems, and more than redeems us. The nation that has the poetry of Chaucer, Spenser, Shakspeare, and Milton, with all the brilliant train down nearly to the present time, must ever hold a distinguished place even as an imaginative people. Shakspeare alone would stamp a language with immortality. The British novels stand high among works of imagination: they have generally had the merit of advancing the cause of morality. Had the French novelists attended more to this, their knowledge of the human heart and the brilliancy of their composition would have been more appreciated.

Poetry of the highest stamp has fled before the utilitarian spirit of the age, yet there is as much talent in the world, and imagination too, at the present time, as ever there was at any period, though directed to different objects; but what is of more importance, there is a constant increase of liberal sentiment and disinterested benevolence. Three of the most beneficial systems of modern times are due to the benevolence of English ladies,—the improvement of prison discipline, savings-banks, and banks for lending small sums to the poor. The success of all has exceeded every expectation at home, and these admirable institutions are now adopted abroad. The importance of popular and agricultural education is becoming an object of attention to the more enlightened governments; and one of the greatest improvements in education is that teachers are now fitted for their duties by being taught the art of teaching. The gentleness with which instruction is conveyed no longer blights the joyous days of youth, but, on the contrary, encourages self-education, which is the most efficient.

The system of infant schools, established in many parts of Europe and throughout the United States of North America, is rapidly improving the moral condition of the people. The instruction given in them is suited to the station of the scholars, and the moral lessons taught are often reflected back on the uneducated parents by their children. Moreover, the personal intercourse with the higher orders, and the kindness which the children receive from them, strengthens the bond of reciprocal good feeling. Since the abolition of the feudal system, the separation between the higher and lower classes of society has been increasing; but the generous exertions of individuals, whose only object is to do good, is now beginning to correct a tendency that, unchecked, might have led to the worst consequences to all ranks.

The voluntary sacrifices that have lately been made to relieve the necessities of a famishing nation show the humane disposition of the age. But it is not one particular and extraordinary case, however admirable, that marks the general progress—it is not in the earthquake or the storm, but in the still small voice of consolation heard in the cabin of the wretched, that is the prominent feature of the charities of the present time, when the benevolent of all ranks seek for distress in the abode of poverty and vice, to aid and to reform. No language can do justice to the merit of those who devote themselves to the reformation of those children who have hitherto wandered neglected in the streets of great cities, in the unpromising task in which they have laboured with patience, undismayed by difficulties that might have discouraged the most determined; but they have succeeded. The language of kindness and sympathy, never before heard by these children of crime and wretchedness, is saving multitudes from perdition. But it would require a volume to enumerate the exertions that are making for the accommodation, health, and improvement of the people, and the devotion of high and low to the introduction of new establishments and the amelioration of the old. Noble and liberal sentiments mark the proceedings of public assemblies, whether in the cause of nations or of individuals; and the severity of our penal laws is mitigated by a milder system. Happily this liberal and benevolent spirit is not confined to Britain; it is universal in the states of the American Union; it is spreading widely through the more civilized countries of Europe. A noble instance that has lately surprised all Europe shows how rapidly the wise measures of a truly great and good sovereign are raising a fine people to that place among the nations which they had lost. No retrograde movement can now take place in civilization; the diffusion of Christian virtues and of knowledge ensures the progressive advancement of man in those high moral and intellectual qualities that constitute his true dignity. But much yet remains to be done at home, especially in religious instruction and the prevention of crime; and abroad millions of our fellow-creatures in both hemispheres are still in the lowest grade of barbarism. Ages and ages must pass away before they can be civilized; but if there be any analogy between the period of man's duration on earth and that of the frailest plant or shell-fish of the geological periods, he must still be in his infancy; and let those who doubt of his indefinite improvement compare the state of Europe in the middle ages, or only fifty years ago, with what it is at present. Some, who seem to have lived before their time, were then prosecuted and punished for opinions which are now sanctioned by the legislature and acknowledged by all. The moral disposition of the age appears in the refinement of conversation. Selfishness and evil passions may possibly ever be found in the human breast; but the progress of the race will consist in the increasing power of public opinion, the collective voice of mankind, regulated by the Christian principles of morality and justice. The individuality of man modifies his opinions and belief; it is a part of that variety which is a universal law of nature; so that there will probably always be difference of views as to religious doctrine, which, however, will become more spiritual and freer from the taint of human infirmity; but the power of the Christian religion will appear in purer conduct, and in the more general practice of mutual forbearance, charity, and love.

THE END.

CPSIA information can be obtained
at www.ICGtesting.com
Printed in the USA
LVHW090101090121
676105LV00004B/40